WIFI FOR THE ENTERPRISE

WiFi for the Enterprise

Nathan J. Muller

McGraw-Hill
New York Chicago San Francisco Lisbon
London Madrid Mexico City Milan New Delhi
San Juan Seoul Singapore Sydney Toronto

The McGraw·Hill Companies

Cataloging-in-Publication Data is on file with the Library of Congress.

1 2 3 4 5 6 7 8 9 0 DOC/DOC 0 9 8 7 6 5 4 3

ISBN 0-07-141252-2

The sponsoring editor for this book was Steve Chapman and the production supervisor was Pamela A. Pelton. It was set in Century Schoolbook by MacAllister Publishing Services, LLC.

Printed and bound by RR Donnelley.

This book is printed on recycled, acid-free paper containing a minimum of 50 percent recycled de-inked fiber.

Dedicated to
Marie & Steve

CONTENTS

PREFACE

Wired networks are here to stay, but wireless networks promise anytime, anywhere access to the Internet or corporate intranets. This is the idea behind *Wireless Fidelity* (WiFi). It provides a convenient way for employees to stay connected to business applications and remain productive, even while they are mobile. The ability to take advantage of corporate resources and the wealth of information on the Internet—regardless of location and without the restrictions of a wired connection—is one of the most compelling new trends of the decade, rivaling e-mail and cell phones in significance. Depending on the equipment selected, users can have wireless connections of up to 11 Mbps in the 2.4 GHz frequency band or up to 54 Mbps in the 5 GHz frequency band. Standards are emerging to boost these speeds even higher, and if the equipment is properly configured, wireless connections can be very secure.

Several compelling needs exist for ubiquitous wireless connections. Workers today need secure access to their organization's network, whether they are at the office, at a branch location, at home, or at an airport. With a wireless connection, they can have full access to e-mail, applications, and data, enabling them to be productive from a variety of locations within range of a wireless *access point* (AP).

In addition, workers require continual access to the Internet and to their organization's private intranet. Wireless technology enables them to take advantage of these capabilities constantly, conducting research and communicating with coworkers and business partners through e-mail and instant messaging, without the physical restrictions of a wired connection.

Finally, most wireless computing is conducted with notebook computers, but that scenario is rapidly changing and broadening. Increasingly, wireless vendors are offering silicon solutions that dramatically reduce the form factor, power consumption, and cost of wireless solutions. Consequently, wireless capabilities are being featured in smaller devices, such as *personal digital assistants* (PDAs) and cell phones. These devices, when linked wirelessly to the company network, are becoming critical productivity tools for workers who cannot spend their entire day sitting at a desk.

What makes WiFi more compelling than any other radio technology, such as *Home Radio Frequency* (HomeRF), Bluetooth, or fixed access services, is its capability to serve multiple markets with equal effectiveness and economy. WiFi is equally suited to the home and enterprise markets, and it creates a viable service that can be offered by telephone companies and *Internet service providers* (ISPs). Its standardization enables users to roam seamlessly between the floors of a building or between buildings in a campus, even allowing users to take work with them to local hot spots in the surrounding community, such as cafés. When traveling, they can pick up a wireless connection while waiting at the airport and finish up with a wireless connection at a hotel or convention center. At this writing, efforts are under way to put together roaming agreements between service providers so that wireless connections can be available virtually anywhere the user happens to be.

Despite WiFi products having been available only since 1999, the technology is already viewed as mature. The marketing efforts of vendors are not so much aimed at convincing individuals and companies to adapt wireless technology as it is convincing user's to adopt a particular brand of equipment.

Much of the market success of this technology can be attributed to the efforts of the WiFi Alliance (formerly the *Wireless Ethernet Compatibility Alliance* [WECA]). The WiFi Alliance is an international nonprofit vendor association formed in 1999 to certify the interoperability of wireless network products based on *Institute of Electrical and Electronics Engineers* (IEEE) 802.11 specifications. It has instituted a test suite that defines how member products are tested to certify that they are interoperable with other WiFi-certified products. These tests, conducted at an independent laboratory, offer buyers the assurance that the products they purchase will work together at a basic level when properly configured. The goal of the WiFi Alliance's members is to enhance the user experience through product interoperability.

Wireless technology for computer connections is not a new idea; the first connections were established two decades ago, but adoption proceeded very slowly for several reasons. The original wireless data rates were too slow to serve mainstream users on a shared *local area network* (LAN). Although throughput did gradually increase, network speeds still lagged behind those of wired LANs. In addition, proprietary solutions dominated the marketplace, providing little interoperability among devices, and denying users the benefits of mixing and matching best-of-breed solutions from a variety of vendors. Finally, these low-speed proprietary solutions were very expensive compared to wired solutions.

In recent years, the situation has changed dramatically. In 1999, the IEEE ratified the 802.11b standard, offering data rates up to 11 Mbps, which is comparable to the 10 Mbps connections in common use for many

Ethernet-based workgroups. For the first time, *wireless LANs* (WLANs) became truly usable for most work environments and office applications. Multiple vendors quickly came to support the 802.11b standard, which drove down costs and increased demand, fueling even greater vendor support. In addition, the 802.11b standard assured users of device interoperability. The WECA was formed to certify the interoperability of WLAN products based on the IEEE 802.11b specification and to promote the use of the standard across all market segments. With the rapid adoption of the 802.11b standard, users began to have a choice in the selection of interoperable, high-performance wireless equipment at an affordable price. Over 500 vendors now offer WiFi products.

Many types of organizations today see a tremendous value in adding wireless links to the corporate LAN. For years, notebook computers have held out the promise of anytime, anywhere computing, but with access to the LAN and the Internet becoming such an integral part of business, a wireless connection is needed to make the promise of anytime, anywhere computing become a reality. Wireless devices enable users to be constantly connected from virtually anywhere: a desk, a conference room, the coffee shop, or another building on a corporate or academic campus. In addition to job autonomy, wireless provides users with maximum flexibility, productivity, and efficiency, while facilitating collaboration and cooperation with colleagues, business partners, and customers. In addition, wireless can bring LAN access to locations where laying cable is difficult or expensive.

As with any technology, wireless is continually evolving, offering advancements in speed, bandwidth, security, and more. As noted, the 802.11b standard, the most widely deployed wireless standard, operates in the 2.4 GHz unlicensed radio band, delivering a maximum data rate of 11 Mbps. For some organizations, this performance is sufficient for their current needs, but others are clamoring for a new generation of devices that will deliver even greater throughput, access, and functionality for their rapidly growing numbers of wireless users.

Wireless vendors are responding. Currently, the IEEE is focusing on two higher-performance standards: 802.11a and 802.11g. The IEEE ratified the 802.11a standard in 1999, but the first 802.11a-compliant products did not begin appearing on the market until December 2001. The 802.11a standard delivers a maximum data rate of 54 Mbps and 8 nonoverlapping frequency channels, resulting in increased network capacity, improved scalability, and the capability to create microcellular deployments without interference from adjacent cells. Operating in the unlicensed portion of the 5 GHz radio band, 802.11a is also immune to interference from devices that operate in the 2.4 GHz band, such as microwave ovens, cordless phones, and Bluetooth (a short-range, low-speed, point-to-point, *personal area network* [PAN]wireless standard).

The 802.11a standard is not, however, compatible with existing 802.11b-compliant wireless devices, but the two technologies can operate in the same vicinity without interfering with each other. Dual-band APs can support both technologies at the same time with separate radio cards. This allows different classes of users to be supported; power users can transfer large files over a high-capacity 5 GHz link, while users of routine office applications can use the 2.4 GHz link. Each class of user will have the bandwidth it needs without causing throughput problems for the other.

A barrier to the widespread adoption of 802.11a has been the lack of interoperability certification. This changed in late 2002, when the WiFi Alliance announced WiFi interoperability certification testing for 5 GHz 802.11a-based products. The 802.11a interoperability testing had been delayed until a basic requirement of the WiFi Alliance had been met: the availability of multiple products based on a second IEEE 802.11a chipset. The announcement of a second chipset was made in April 2002 and products were available for testing two months later. The availability of products allowed the WiFi Alliance to develop the interoperability benchmark against which all products are tested. As with 802.11b products, 802.11a products that pass the certification will be granted the WiFi seal of interoperability.

Another standard, 802.11g, delivers the same 54 Mbps maximum data rate as 802.11a, but with an additional and compelling advantage: backward compatibility with 802.11b equipment. This means that 802.11b client cards will work with 802.11g APs and 802.11g client cards will work with 802.11b APs. Because 802.11g and 802.11b operate in the same 2.4 GHz unlicensed band, migrating to 802.11g will be an affordable choice for organizations with existing 802.11b wireless infrastructures. Much like Ethernet at 10 Mbps and Fast Ethernet at 100 Mbps, 802.11g products can be mixed and matched with 802.11b products on the same network.

The wireless wave is just beginning. The new emphasis on worker mobility will drive the unprecedented growth of WLANs worldwide. Organizations of all types and sizes, therefore, must begin planning their wireless strategies today. Wireless technologies and services have become so popular worldwide, as well as sufficiently sophisticated and complex, as to merit dozens of books on the topic being published every year. This book clearly explains the essential concepts of wireless, including services, applications, protocols, administration and management, security, and deployment issues.

The information contained in this book, especially as it relates to specific vendors and products, is believed to be accurate at the time it was written and is, of course, subject to change with continued advancements in technologies and shifts in market forces. Any mention of specific products and services is for illustration purposes only and does not constitute an endorsement of any kind by either the author or the publisher.

Nathan J. Muller

CHAPTER 1

WiFi in Perspective

Of all the communications services available today, wireless services are having the most dramatic impact on our personal and professional lives, enhancing personal productivity, mobility, and security. In particular, the impact of cellular phone services on our lives is well documented, but *Wireless Fidelity* (WiFi) also promises to have a dramatic effect in the near future. In fact, emerging broadband cellular phone and WiFi services are not mutually exclusive, but complementary, so much so that a single PC Card for notebooks and some *personal digital assistants* (PDAs) will soon support both services, switching between the two networks automatically as the user changes locations or applications.

WiFi operates in unlicensed frequency bands and is based on a set of standards promulgated by the *Institute of Electrical and Electronics Engineers* (IEEE). One standard, called 802.11b, specifies the requirements for connecting devices at the maximum throughput rate of 11 Mbps using the 2.4 GHz frequency band, whereas 802.11a specifies the requirements for connecting devices at the maximum throughput rate of 54 Mbps using the 5 GHz frequency band. Proprietary extensions to each of these standards enable speed bursts of 22 Mbps and 72 Mbps, respectively.

As the cost of wireless equipment continually decreases to the point of reaching parity with wired gear, WiFi networks are now being used in a number of settings, such as college campuses, business parks, office buildings, and even homes. Such networks are also being implemented by a number of service providers in public places such as airports, hotels, retail locations, and cafes to give users of notebook computers and handheld devices wireless access to the Internet for e-mail and web browsing. In the corporate environment, WiFi enables users to access *local area networks* (LANs) to search databases, share files, and print documents—all without requiring them to find an available port and set up a cable connection. And since many employees visit other locations in a building or campus throughout the day, wireless connections facilitate mobility without impeding productivity.

Aside from the low cost of equipment, the growing popularity of WiFi networks has occurred for several other reasons:

- They not only work, but they work well, and they are undergoing continuous refinement, particularly in the area of security.
- Wireless connections are easy to set up, especially with Windows XP, which provides integral support for WiFi, eliminating the need to manually install drivers. Some notebook computers even come with WiFi antennas embedded into their lids, eliminating the need for a PC Card for *wireless LAN* (WLAN) connections.

- There is nothing new to learn about using WiFi; anyone who uses an Ethernet LAN at work or home will readily appreciate the convenience and performance of WiFi, which is also based on Ethernet.
- Connectivity is available from a growing number of service providers, so WiFi can be used between the home and workplace at various *hot spots* such as airports, hotels, and cafes, which greatly extends its utility.

Many other wireless technologies are available. To put WiFi into the proper context, it is helpful to survey some of the other wireless alternatives available to businesses and telecommuters, and examine the applications for which they are best suited. Sometimes WiFi will complement another wireless technology such as Bluetooth or *General Packet Radio Service* (GPRS), enabling the user to benefit from having access to both, depending on the location or type of application.

Bluetooth

Bluetooth is an omnidirectional wireless technology that provides limited range voice and data transmission over the same unlicensed 2.4 GHz frequency band used by WiFi, allowing connections with a wide variety of fixed and portable devices that normally would have to be cabled together. Up to eight devices—one master and seven slaves—can communicate with one another in a so-called piconet at distances of up to 30 feet. Table 1-1 summarizes the performance characteristics of Bluetooth.

Applications

Among other things, users can swap data and synchronize files using Bluetooth merely by having the devices come within range of one another. Images captured with a digital camera, for example, can be dropped off at a PC for editing or a color printer for output on photo-quality paper—all without having to connect cables, load files, open applications, or click buttons.

The technology is a combination of circuit switching and packet switching, making it suitable for voice as well as data. Instead of fumbling with a cell phone while driving, for example, a user can wear a lightweight headset to answer a call and engage in a conversation even if the phone is tucked away in a briefcase or purse.

Table 1-1

Performance characteristics of Bluetooth

Feature/Function	Performance
Frequency band	2.4 GHz *Industrial, Scientific, and Medical* (ISM) band.
Connection type	*Frequency-hopping spread spectrum* (FHSS).
Hop rate	1,600 hops per second among 79 frequencies.
Transmission power	1 *milliwatt* (mW).
Aggregate data rate	1 Mbps using frequency hopping.
Range	Up to 30 feet (9 meters).
Supported stations	Up to 8 devices per piconet.
Voice channels	Up to 3 synchronous channels.
Data security	For authentication, a 128-bit key is used; for encryption, the key size is configurable between 8 and 128 bits.
Addressing	Each device has a 48-bit *Media Access Control* (MAC) address that is used to establish a connection with another device.

Although WLANs are useful in minimizing the need for cables, they are not normally used for interconnecting the range of mobile devices people carry around everyday between home and the office, such as smartphones, PDAs, MP3 players, and digital cameras. For this, Bluetooth is needed. In the office, a Bluetooth portable device can be in motion while connected to the LAN *access point* (AP) as long as the user stays within the 30-foot range.

Bluetooth can be combined with other technologies to offer wholly new capabilities, such as automatically lowering the ring volume of a cell phone or shutting it off as a user enters quiet zones like churches, restaurants, theaters, and classrooms. Upon leaving the quiet zone, the cell phone is returned to the original settings.

Topology

Bluetooth devices within a piconet play one of two roles: master or slave. The master is the device in a piconet whose clock and hopping sequence is used to synchronize all other devices (slaves) in the piconet. The unit that carries out the paging procedure and establishes a connection is the master of the connection by default. The slaves are the units within a piconet that are synchronized to the master via its clock and hopping sequence.

The Bluetooth topology is best described as a multiple piconet structure. Because Bluetooth supports both point-to-point and point-to-multipoint connections, several piconets can be established and linked together in a topology called a *scatternet* whenever the need arises (see Figure 1-1).

Piconets are uncoordinated, with frequency hopping occurring independently. Several piconets can be established and linked together ad hoc, where each piconet is identified by a different frequency-hopping sequence. All users participating on the same piconet are synchronized to this hopping sequence. Although the synchronization of different piconets is not permitted in the unlicensed ISM band, Bluetooth units may participate in different piconets through *time division multiplexing* (TDM). This enables a unit to sequentially participate in different piconets by being active in only one piconet at a time.

With its service discovery protocol, Bluetooth enables a much broader vision of networking, including the creation of *personal area networks* (PANs), where all the devices in a person's life can communicate and work together. Technical safeguards ensure that a cluster of Bluetooth devices in public places, such as an airport lounge or train terminal, would not suddenly start talking to one another.

Figure 1-1 The possible topologies of networked Bluetooth devices, where each is either a master or slave

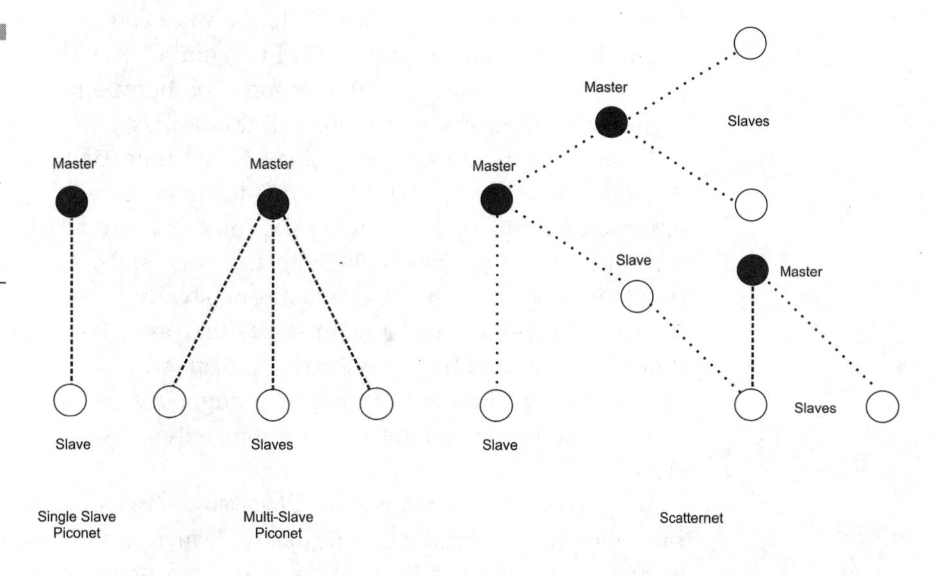

Technology

Two types of links have been defined for Bluetooth in support of voice and data applications: an *asynchronous connectionless* (ACL) link and a *synchronous connection-oriented* (SCO) link. ACL links support data traffic on a best-effort basis. The information carried can be user data or control data. SCO links support real-time voice and multimedia traffic using reserved bandwidth. Both data and voice are carried in the form of packets, and Bluetooth devices can support active ACL and SCO links at the same time.

ACL links support symmetrical/asymmetrical, packet-switched, point-to-multipoint connections that are typically used for data. For symmetrical connections, the maximum data rate is 433.9 Kbps in both directions—send and receive. For asymmetrical connections, the maximum data rate is 723.2 Kbps in one direction and 57.6 Kbps in the reverse direction. If errors are detected at the receiving device, a notification is sent in the header of the return packet, indicating that only lost or corrupt packets need to be retransmitted.

SCO links provide symmetrical, circuit-switched, point-to-point connections that are typically used for voice. Three synchronous channels of 64 Kbps each are available for voice. The channels are derived through the use of either *pulse code modulation* (PCM) or *continuous variable slope delta* (CVSD) modulation. PCM is the standard for encoding speech in analog form into the digital format of ones and zeros. CVSD is another standard for analog-to-digital encoding, but offers more immunity to interference and therefore is better suited than PCM for voice communication over a wireless link. Bluetooth supports both PCM and CVSD; the appropriate voice-coding scheme is selected after negotiation between the link managers of each Bluetooth device before the call takes place.

Voice and data are sent as packets. Communication is handled with *time division duplexing* (TDD), which divides the channel into time slots, each 625 *microseconds* (μs) in length. The time slots are numbered according to the clock of the piconet master. In the time slots, master and slave can transmit packets. In the TDD scheme, master and slave alternatively transmit (see Figure 1-2). The master starts its transmission in even-numbered time slots only, and the slave starts its transmission in odd-numbered time slots only. The start of the packet is aligned with the slot start. Packets transmitted by the master or slave may extend over as many as five time slots.

With TDD, bandwidth can be allocated on an as-needed basis, changing the makeup of the traffic flow as demand warrants. For example, if the user wants to download a large data file, the amount of bandwidth that is

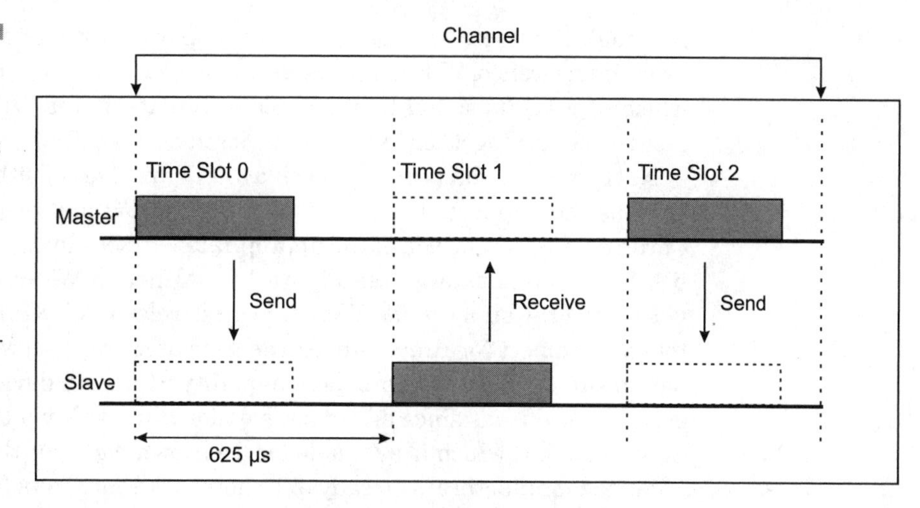

Figure 1-2 With the TDD scheme used in Bluetooth, packets are sent over time slots of 625 μs in length between the master and slave units within a piconet.

needed will be allocated to the transfer. Then, at the next moment, if a file is being uploaded, that same amount of bandwidth can be allocated to that transfer.

Regardless of the application—voice or data—making connections between Bluetooth devices is as easy as powering them up. In fact, one advantage of Bluetooth is that it does not need to be set up—it is always on, running in the background and looking for other devices it can communicate with. When Bluetooth devices come within range of one another, they engage in a service discovery procedure, which entails the exchange of messages to become aware of each other's service and feature capabilities. Having located available services within the vicinity, the user may select from any of them. After that, a connection between two or more Bluetooth devices can be established.

The radio link itself is very robust, using FHSS technology to overcome interference and fading. Spread spectrum is a digital coding technique in which the signal is taken apart, or spread, so that it sounds more like noise as it is sent through the air. The addition of frequency hopping—having the signals skip from one frequency to another—makes wireless transmissions even more secure. Bluetooth specifies a rate of 1,600 hops per second among 79 frequencies. Because only the sender and receiver know the hopping sequence for coding and decoding the signal, eavesdropping is virtually impossible. For enhanced security, Bluetooth also supports device authentication and encryption.

Other frequency-hopping transmitters in the vicinity will use different hopping patterns and much slower hop rates than Bluetooth devices.

Although Bluetooth signals constantly hop over a range of frequencies to avoid interference, WiFi employs *direct sequence spread spectrum* (DSSS), which exposes its signal to interference. Bluetooth and WiFi would not be able to operate together in the same vicinity.

Nevertheless, notebook users can avail themselves of either Bluetooth or WiFi as the application may warrant. It is just a matter of swapping PC Cards to take advantage of the appropriate wireless link, which will establish itself automatically (see Figure 1-3). Although Windows XP now provides integral support for WiFi, the first release of XP did not provide integral support for Bluetooth. At the time of its release, Microsoft did not have a sufficient array of production-quality Bluetooth devices to test. However, Microsoft has since added support for Bluetooth via the Windows XP Service Pack 1, which is available for free download over the Internet.

Some manufacturers already offer notebook computers for the corporate market that have both Bluetooth and WiFi antennas embedded into their lids. The Tecra 9100 from Toshiba Computer Systems Group, for example, integrally supports Bluetooth and WiFi, enabling the user to save the vacant PC Card slot for other purposes. With 2.5G cellular services becoming widely available, offering up to 144 Kbps for data applications, the slot could be used for a GPRS card. All this, plus the notebook's standard *infrared* (IR) port, turns the computer into a quad-mode wireless device.

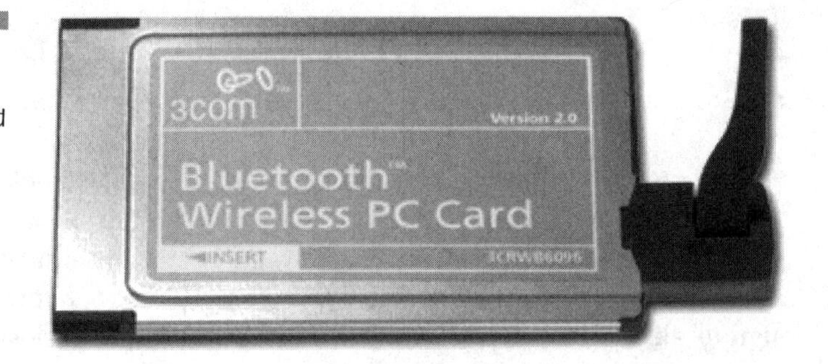

Figure 1-3 3Com's Wireless Bluetooth PC Card fits into a notebook's PC Card slot, letting the user communicate with other Bluetooth products, including PDAs, printers, digital cameras, and other enabled computers.

Infrared

In addition to its use as a wireless interface to connect notebooks and other portable devices to the desktop computer, infrared technology can be used to implement WLANs over the wavelength band between 780 and 950 *nanometers* (nm). Two categories of infrared systems are commonly used for WLANs. One is *directed infrared*, which uses a very narrow laser beam to transmit data over one to three miles. This approach can be used to connect LANs in different buildings.

The other category is *nondirected infrared*, which uses a less focused approach. Instead of using a narrow beam to convey the signal, the light energy is spread out and bounced off narrowly defined target areas or larger surfaces such as office walls and ceilings. Nondirected infrared links may be further categorized as either line of sight or diffuse (see Figure 1-4). Line-of-sight links require a clear path between the transmitter and receiver, but generally offer higher performance.

The line-of-sight limitation may be overcome by incorporating a recovery mechanism in the infrared LAN, which is managed and implemented by a separate device called a *multiple access unit* (MAU) to which each workstation is connected. When a line-of-sight signal between two stations is temporarily blocked, the MAU's internal optical link control circuitry automatically changes the link's path to get around the obstruction. When the original path is cleared, the MAU restores the link over that path. No data is lost during this recovery process.

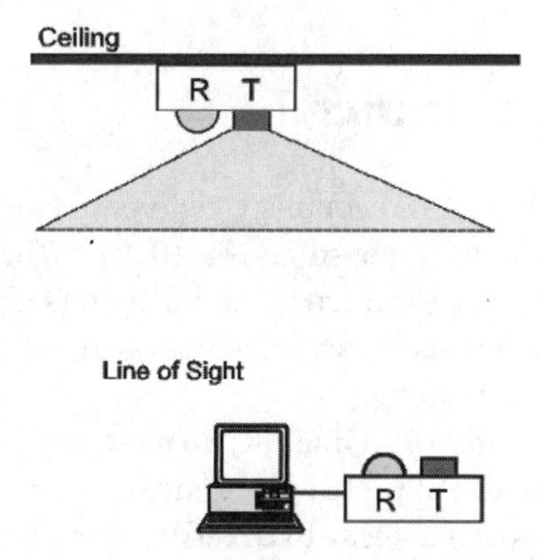

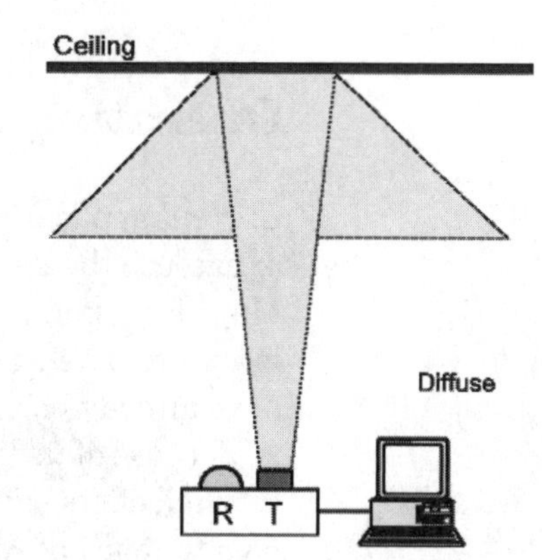

Figure 1-4 Line-of-sight versus diffuse configurations for infrared links

Diffuse links rely on light bounced off reflective surfaces. Because it is difficult to block all of the light reflected from large surface areas, diffuse links are generally more robust than line-of-sight links. The disadvantage of diffused infrared is that a great deal of energy is lost, and, consequently, the data rates and operating distances are much lower.

System Components

Light-emitting diodes (LEDs) or *laser diodes* (LDs) are used for transmitters. LEDs are less efficient than LDs. They typically exhibit only 10 to 20 percent electro-optical power conversion efficiency, whereas LDs offer 30 to 70 percent electro-optical conversion efficiency. However, LEDs are much less expensive than LDs, which is why most commercial systems use them.

Two types of low-capacitance silicon photodiodes are used for receivers: *positive intrinsic negative* (PIN) and avalanche. The simpler and less expensive PIN photodiode is typically used in receivers that operate in environments with bright illumination, whereas the more complex and expensive avalanche photodiode is used in receivers that must operate in environments where background illumination is weak. The difference in the two types of photodiodes is their sensitivity.

The PIN photodiode produces an electrical current in proportion to the amount of light energy projected onto it. Although the avalanche photodiode requires more complex receiver circuitry, it operates in much the same way as the PIN diode, except that when light is projected onto it, a slight amplification of the light energy occurs. This makes it more appropriate for weakly illuminated environments. The avalanche photodiode also offers a faster response time than the PIN photodiode.

Operating Performance

Current applications of infrared technology yield performance that matches or exceeds the data rate of wire-based LANs: 10 Mbps for Ethernet and 16 Mbps for token ring. However, infrared technology has a much higher performance potential—transmission systems operating at 50 and 100 Mbps have already been demonstrated.

Because of its limited range and inability to penetrate walls, nondirected infrared offers some measure of protection against eavesdropping. Even signals that go out windows are useless to eavesdroppers because they do not travel far and may be distorted by impurities in the glass as well as by the glass placement angle.

Infrared offers high immunity from *electromagnetic interference* (EMI), which makes it suitable for operation in harsh environments like factory floors. Because of its limited range and inability to penetrate walls, several infrared LANs may operate in different areas of the same building without interfering with each other. Because there is less chance of multipath fading (large fluctuations in the received signal amplitude and phase), infrared links are highly robust.

Many indoor environments have incandescent or fluorescent lighting, which induces noise in infrared receivers. This is overcome by using directional infrared transceivers with special filters to reject background light.

Media Access Control (MAC)

Infrared supports both contention-based and deterministic MAC techniques, making it suitable for Ethernet as well as token ring LANs.

To implement Ethernet's contention protocol, *carrier sense multiple access* (CSMA), each computer's infrared transceiver is typically aimed at the ceiling. Light bounces off the reflector in all directions to enable each user to receive data from other users (see Figure 1-5). CSMA ensures that only one station can transmit data at a time. Only the station(s) to which packets are addressed can actually receive them.

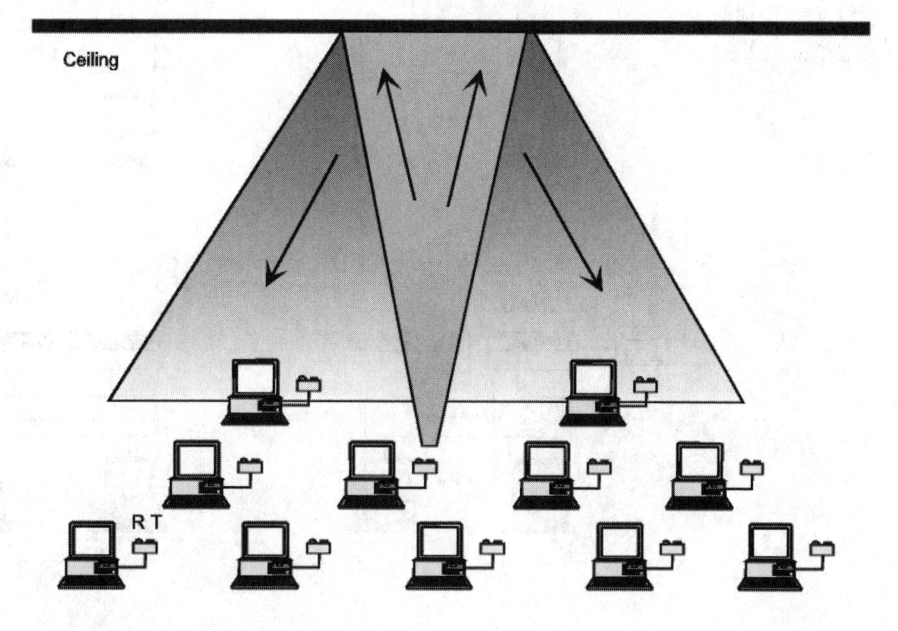

Figure 1-5 The implementation of Ethernet using diffuse infrared

Deterministic MAC relies on token passing to ensure that all stations in turn have an equal chance to transmit data. This technique is used in token ring LANs, where each station uses a pair of highly directive (line-of-sight) infrared transceivers. The outgoing transducer is pointed at the incoming transducer of a station down line, thus forming a closed ring with the wireless infrared links among the computers (see Figure 1-6). With this configuration, much higher data rates can be achieved because of the gain associated with the directive infrared signals. This approach improves the overall throughput, since fewer bit errors will occur, which minimizes the need for retransmissions.

Infrared Computer Connectivity

Most notebook computers and PDAs have infrared ports, and just about every major mobile phone brand includes at least one infrared-enabled handset. Infrared products for computer connectivity conform to the standards developed by the *Infrared Data Association* (IrDA). The standard

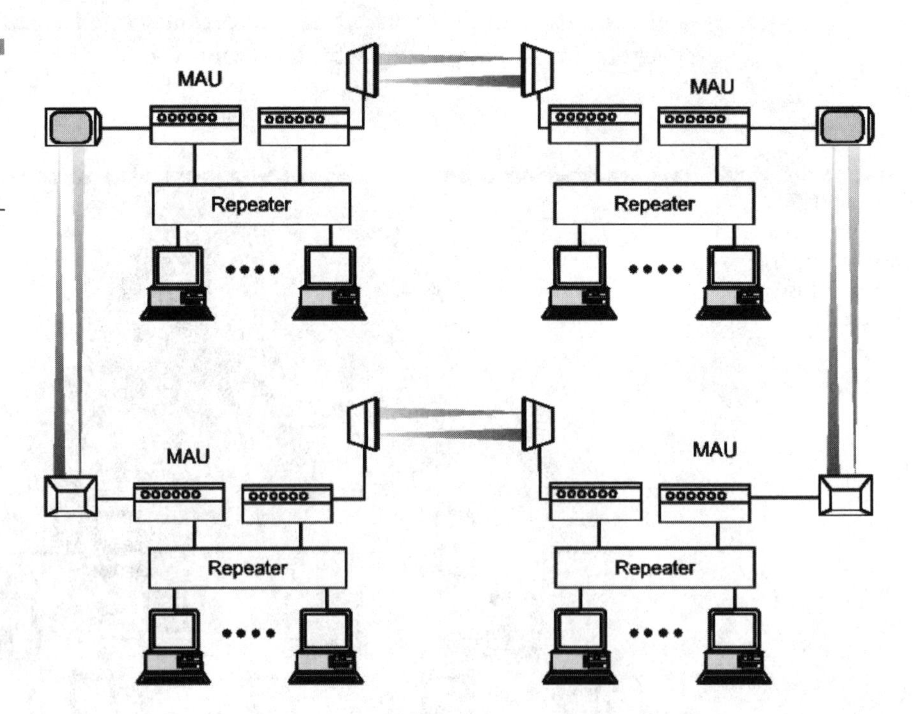

Figure 1-6 The implementation of token ring using line-of-sight infrared

protocols include *Serial Infrared* (SIR) at 115 Kbps, *Fast Infrared* (FIR) at 4 Mbps, and *Very Fast Infrared* (VFIR) at 16 Mbps. The higher speed available with VFIR is intended to address the new demands of transferring large image files between digital cameras, scanners, and PCs. Table 1-2 summarizes the performance characteristics of the IrDA's infrared standard.

Infrared's primary impact will take the form of benefits for mobile professional users. It enables simple point-and-shoot connectivity between devices to enable users to exchange information and reap more of the productivity gains promised by portable computing. IrDA technology is supported in over 100 million electronic devices including desktop, notebook, and palm PCs; printers; digital cameras; public phones/kiosks; cellular phones; pagers; PDAs; electronic books; electronic wallets; and other mobile devices.

When used on a LAN, infrared technology also confers substantial benefits to network administrators. Infrared is easy to install and configure, requires no maintenance, and imposes no remote-access tracking hassles. It does not disrupt other network operations, and it provides data security. Because it makes connectivity so easy, it encourages the use of high-productivity network and groupware applications, thus helping administrators amortize the costs of these packages across a larger user base.

Table 1-2

Performance characteristics of the IrDA's infrared standard

Feature/Function	Performance
Connection type	Infrared, narrow beam (30° angle or less).
Spectrum	Optical, 850 nm.
Transmission power	100 mW.
Data rate	Up to 16 Mbps using VFIR.
Range	Up to 3 feet (1 meter).
Supported devices	Two.
Data security	The short range and narrow angle of the infrared beam provide a simple form of security; otherwise, no security capabilities exist at the link level.
Addressing	Each device has a 32-bit physical ID that is used to establish a connection with another device.

Points of Convergence

In some ways, Bluetooth competes with infrared; in other ways, the two technologies are complementary. With both infrared and Bluetooth, data exchange is considered to be a fundamental function. Data exchange can be as simple as transferring business card information from a mobile phone to a palmtop, or as sophisticated as synchronizing personal information between a palmtop and desktop PC. In fact, both technologies can support many of the same applications, which raises the following question: Why would users need both technologies?

The answer lies in the fact that each technology has advantages and disadvantages. The very scenarios that leave infrared falling short are the ones where Bluetooth excels, and vice versa. Take the electronic exchange of business card information between two devices. This application will usually take place in a conference room or exhibit floor where a number of other devices might be attempting to do the same thing. Infrared excels in this situation. The short range and narrow angle of infrared—30° or less—enable each user to aim his or her device at the intended recipient with point-and-shoot ease. Close proximity to another person is natural in a business card exchange situation, as is pointing one device at another. The limited range and narrow angle of infrared enable other users to perform a similar activity with ample security and no interference.

In the same situation, a Bluetooth device would not perform as well as an infrared device. With its omnidirectional capability, the Bluetooth device must first discover the intended recipient. The user cannot simply point at the intended recipient—a Bluetooth device must perform a discovery operation that will probably reveal several other Bluetooth devices within range, so close proximity offers no advantage here. The user will be forced to select from a list of discovered devices and apply a security mechanism to prevent unauthorized access. All this makes the use of Bluetooth for business card exchange an awkward and needlessly time-consuming process.

In other data exchange situations, however, Bluetooth might be the preferred choice. Bluetooth's capability to penetrate solid objects and communicate with other devices in a piconet allows for data exchange opportunities that are very difficult or impossible with infrared. For example, Bluetooth enables a user to synchronize a mobile phone with a notebook computer without taking the phone out of the user's jacket pocket or purse. This enables the user to type a new address at the computer and move it to the mobile phone's directory without unpacking the phone and setting up a cable connection between the two devices. The omnidirectional capability of Bluetooth enables synchronization to occur instantly, assuming that the phone and computer are within 30 feet of each other.

Using Bluetooth for synchronization does not require that the phone remain in a fixed location. If a phone is carried about in a briefcase, the synchronization can occur while the user moves around. This is not possible with infrared because the signal is not able to penetrate solid objects and the devices must be within a few feet of each other. Furthermore, the use of infrared requires that both devices remain stationary while the synchronization occurs.

When it comes to data transfers, infrared does offer a significant speed advantage over Bluetooth. Whereas Bluetooth moves data between devices at an aggregate rate of 1 Mbps, infrared offers up to 16 Mbps. Even when Bluetooth is enhanced for higher data rates in the future, infrared is likely to maintain its speed advantage for many years to come.

Bluetooth complements infrared's point-and-shoot ease of use with omnidirectional signaling, longer-distance communications, and the capacity to penetrate walls. For some users, having both Bluetooth and infrared will provide the optimal short-range wireless solution. For others, the choice of adding Bluetooth or infrared will be based on the applications and intended usage.

Home Radio Frequency (HomeRF)

Telecommuters are becoming increasingly interested in connecting computers and peripherals at home, possibly tying in their notebooks from work as well, so they can access the Internet or corporate intranet from a shared broadband connection like *Digital Subscriber Line* (DSL) or cable. One method of implementing a wireless network in the home is to use products that adhere to the standards of the *Home Radio Frequency Working Group* (HomeRF WG).

HomeRF is positioned as a global extension of *Digitally Enhanced Cordless Telephony* (DECT), the popular cordless phone standard that enables different brands to work together so certified handsets from one vendor can communicate with base stations from another. DECT has been largely confined to Europe because its native 1.9 GHz frequency band requires a license elsewhere, but HomeRF extends DECT to other regions by using the license-free 2.4 GHz frequency band, which is also used by Bluetooth and

WiFi. HomeRF also adds functionality by blending several industry standards, including IEEE 802.11 FHSS for data and DECT for voice.

The standards for HomeRF are addressed by a consortium of vendors called the HomeRF WG. Under the HomeRF standard, compliant devices carry both voice and data traffic and interoperate with the *Public Switched Telephone Network* (PSTN) and the Internet. The standard specifies the use of *Time Division Multiple Access* (TDMA) to provide the delivery of interactive voice and other time-critical services, as well as *carrier sense multiple access / collision avoidance* (CSMA/CA) for the delivery of high-speed packet data. Table 1-3 summarizes the main characteristics of HomeRF.

Applications

The HomeRF standard provides the basis for a broad range of home networking applications, including the following:

- Shared access to the Internet from anywhere in the home, enabling a user to browse the Web from a notebook on the deck or have stock quotes delivered to a PC in the den

Table 1-3
HomeRF characteristics

Feature/Function	Performance
Frequency-hopping network	50 hops per second
Frequency range	2.4 GHz ISM band
Transmission power	100 mW
Data rate	1.6 Mbps with HomeRF 1.0 10 Mbps with HomeRF 2.0 25 Mbps with HomeRF 3.0 (future)
Range	Covers up to 150 feet for typical home and yard
Total network devices	Up to 127
Voice connections	Up to 4 active handsets
Data security	Blowfish encryption algorithm (over 1 trillion codes)
Data compression	LZRW3-A algorithm
48-bit network ID	Enables concurrent operation of multiple co-located networks

- Automatic intelligent routing of incoming telephone calls to one or more cordless handsets, fax machines, or voice mailboxes of individual family members
- Cordless handset access to an integrated message system to review stored voice mail, faxes, and e-mail
- WLANs, enabling users to share files and peripherals between one or more PCs, no matter where they are located within the home
- Spontaneous control of security, electrical, heating, and air conditioning systems from anywhere in or around the home
- Multiuser computer games playable in the same room or in multiple rooms throughout the home

Network Topology

The HomeRF system can operate either as an ad hoc network or as a managed network under the control of a *connection point*. In an ad hoc network, where only data communication is supported, all stations are equal and control of the network is distributed between the stations. For time-critical communications such as interactive voice, a *connection point* is required to coordinate the system. The connection point, which provides the gateway to the PSTN, can be connected to a PC via a standard interface that will enable enhanced voice and data services such as the *universal serial bus* (USB). The HomeRF system also can use the connection point to support power management for prolonged battery life by scheduling device wakeup and polling. The network can accommodate a maximum of 127 nodes. The nodes consist of four basic types:

- A connection point that supports voice and data services
- A voice terminal that only uses the TDMA service to communicate with a base station
- A data node that uses the CSMA/CA service to communicate with a base station and other data nodes
- An integrated node that can use both TDMA and CSMA/CA services

HomeRF uses intelligent hopping algorithms that detect wideband, static interference from microwave ovens, cordless phones, baby monitors, and WiFi networks. Once detected, the HomeRF hop algorithm adapts so no two consecutive hops occur within this interference range. This means that, with very high probability, a packet lost due to interference will get through when it retries on the next hop. Although these algorithms benefit data applications, they are especially important for voice, which requires extremely low *bit error rates* (BER) and low latency. WiFi does not use the frequency-hopping mechanism; instead, it uses DSSS. In doing so, however, WiFi does not support real-time applications such as voice, and it is not very resistant to interference from other household devices. Standards currently under development for WiFi are addressing these issues.

Future Plans

Work has already begun on the future HomeRF 2.1 specification, which will add features designed to reinforce its advantages for voice. Planned enhancements also will enable HomeRF to run WiFi, leading to a peaceful coexistence between the two and giving users the added functionality of HomeRF.

HomeRF 2.0 already supports up to eight phone lines, eight registered handsets, and four active handsets with voice quality and range comparable to leading 2.4 GHz phone systems. With that many lines, each family member can have a personal phone number. HomeRF 2.1 plans to increase the number of active handsets with the same or better voice quality, thus supporting the needs of small businesses.

The 150-foot range of HomeRF already covers most homes into the yard. HomeRF 2.1 will extend that range for larger homes and businesses by using wireless repeaters that are similar to enterprise access points but without the need to connect each one to Ethernet. HomeRF frequency-hopping technology also avoids the complexity of assigning *radio frequency* (RF) channels to multiple APs (or repeaters), and offers easy and effective security and interference immunity. This is especially important because households and small businesses do not usually have network administrators.

To enable individuals to roam across very large homes and fairly large offices while talking on the phone and without losing their voice connection, HomeRF 2.1 will also support voice roaming with soft handoff between repeaters.

HomeRF 2.0 supports Ethernet speeds up to 10 Mbps with fallback speeds and backward compatibility to earlier versions of HomeRF. Performance can be further enhanced to about 20 Mbps. The HomeRF WG is evaluating the need for such enhancements at 2.4 GHz in light of its planned support of WiFi at 5 GHz.

A proposed change to the *Federal Communications Commission's* (FCC's) Part 15 rules governing the 2.4 GHz ISM band will allow adaptive frequency hopping. Although these proposed techniques are not legal today, they enable hoppers such as Bluetooth and HomeRF to recognize and avoid interference from static frequency technologies such as WiFi. Because HomeRF already adjusts its hopping pattern based on interference to ensure that two consecutive hops do not land on interference, supporting this FCC proposal seems trivial.

The HomeRF WG believes in the peaceful coexistence of 2.4 and 5 GHz since each frequency band and technology has specific strengths that complement each other. Rather than draft a specification for 5 GHz, the group simply endorses 802.11a (also known as *WiFi5*) for high-bandwidth applications such as high-definition video streaming and MPEG2 compression. It plans to write application briefs describing how to bridge between 2.4 and 5 GHz technologies, including how to handle differences in *quality of service* (QoS).

Home users have a need for a wireless network that is easy to use, cost effective, and spontaneously accessible. It should also be able to carry voice and data communications. Certified HomeRF products are available today from consumer brands such as Compaq, Intel, Motorola, Proxim, and Siemens through retail, online, and service provider channels. They come in a variety of form factors such as USB and PC Card adapters, residential gateways, and a growing variety of devices that embed HomeRF.

Windows XP does not include integral support for HomeRF. Instead, Microsoft chose to have XP integrally support WiFi, which many industry analysts believe will eventually overtake HomeRF. Even Intel, an early supporter of HomeRF, is now focused exclusively on WiFi. Furthermore, with the growing popularity of Wi-Fi, it is unlikely that users will want to have two technologies—one for the office and another for home. With no appreciable price difference between the two, telecommuters are more likely to standardize on the technology they use in the workplace.

WiFi

As noted earlier, WiFi refers to a version of Ethernet specified under the IEEE 802.11a and 802.11b standards for LANs operating in the 5 GHz and 2.4 GHz unlicensed frequency bands, respectively. WiFi is equally suited to businesses of all types and sizes as well as their telecommuting and remote employees working at home or in branch offices. Equipment is available that enables both bands to be used to support separate networks simultaneously. Some APs, for example, come with dual slots for 2.4 and 5 GHz radio cards, supporting devices on both networks at the same time.

The 802.11 standard makes the wireless network a straightforward extension of the wired network. This has allowed for a very simple implementation of wireless communication with obvious benefits—these can be installed using the existing network infrastructure with minimal retraining or system changes. Notebook users can work anywhere in a building or campus while remaining in contact with the network via strategically placed APs that are plugged into the wired network. Likewise, PDA users can roam throughout the workplace and stay in contact with the corporate network via the same APs, giving them high-speed access to the Internet, e-mail, and network resources. Users can also hot sync their data as they move about, so their information is always up-to-date.

Wireless users can run the same network applications they use on an Ethernet LAN. For most users, no noticeable functional difference exists between a wired Ethernet desktop computer and a wireless computer equipped with a wireless adapter other than the added benefit of the ability to roam within the wireless cell. Under many circumstances, it may be desirable for mobile network devices to link to a conventional Ethernet LAN in order to use servers, printers, or an Internet connection supplied through the wired LAN. A wireless AP is one device that can be used to provide this link.

The IEEE 802.11b standard designates devices that operate in the 2.4 GHz band to provide a data rate of up to 11 Mbps at a range of up to 300 feet (100 meters) indoors and 1,800 feet (600 meters) outdoors using DSSS technology. But with high-gain, line-of-sight antennas, a range of up to 50 miles is possible. Some vendors have implemented proprietary extensions to the 802.11b standard, allowing applications to burst beyond 11 Mbps to reach as much as 22 Mbps. Users can share files and applications, exchange e-mail, access printers, share access to the Internet, and perform any other task as if they were directly cabled to the network.

The IEEE 802.11a standard designates devices that operate in the 5 GHz band to provide a data rate of up to 54 Mbps at a range of up to 900 feet (300 meters) indoors using DSSS technology. Sometimes called WiFi5, this amount of bandwidth enables users to transfer large files quickly or even watch a movie in MPEG format over the network without noticeable delays. This technology works by transmitting high-speed digital data over a radio wave utilizing *orthogonal frequency division multiplexing* (OFDM) technology.

OFDM works by splitting the radio signal into multiple smaller subsignals, which are then transmitted simultaneously at different frequencies to the receiver. OFDM reduces the amount of interference in signal transmissions, which results in a high-quality connection. WiFi5 products automatically sense the best possible connection speed to ensure the greatest speed and range possible with the technology. Some vendors have implemented proprietary extensions to the 802.11a standard, enabling applications to burst beyond 54 Mbps to reach as much as 72 Mbps.

WiFi networks can be implemented in infrastructure mode or ad hoc mode. In infrastructure mode—referred to in the IEEE specification as the *basic service set*—each wireless client computer associates with an AP via a radio link. The AP connects to the 10/100 Mbps Ethernet enterprise network using a standard Ethernet cable and provides the wireless client computer with access to the wired Ethernet network. Ad hoc mode is the peer-to-peer network mode, which is suitable for very small installations. Ad hoc mode is referred to in the 802.11b specification as the *independent basic service set*.

Security for WiFi networks is handled by the IEEE standard called *Wired Equivalent Privacy* (WEP), which is commonly available in 64- and 128-bit versions. The more bits in the encryption key, the more difficult it is for hackers to decode the data. It was originally believed that 128-bit encryption would be virtually impossible to break due to the large number of possible encryption keys. However, hackers have since developed methods to break 128-bit WEP without having to try each key combination, proving that this system is not totally secure. These methods are based upon the ability to gather enough packets off the network using special eavesdropping equipment to determine the encryption key. Although WEP can be broken, it does take considerable effort and expertise to do so. To help thwart hackers, WEP should be enabled on all wireless devices and the keys should be rotated on a frequent basis.

The WLAN industry has recognized that WEP is not as secure as it was once thought and is responding by developing another standard, known as 802.11i, which will enable WEP to use the *Advanced Encryption Algorithm* (AES) to make the encryption key even more difficult to determine. AES replaces the older 56-bit *Digital Encryption Standard* (DES), which had been in use since the 1970s. AES can be implemented in 128-, 192-, and 256-bit versions. For a computer with enough processing power to test 255 keys per second, it would take 149 trillion years to crack AES.

WiFi is a certification of interoperability awarded by the WiFi Alliance, formerly known as the *Wireless Ethernet Compatibility Alliance* (WECA). The WiFi seal indicates that a device has passed independent tests and will reliably interoperate with all other WiFi-certified equipment. Customers benefit from this standard by avoiding becoming locked into one vendor's solution—they can purchase WiFi-certified AP and client devices from different vendors and still expect them to work together.

Table 1-4
WiFi characteristics

Feature/Function	Performance
Frequency range	2.4 and 5 GHz ISM band.
Transmission power	100 mW.
Data rate	11 Mbps at 2.4 GHz and 54 Mbps at 5 GHz. 54 Mbps at 2.4 GHz under 802.11g (future).
Range	2.4 GHz systems: indoors at up to 300 feet (100 meters) from the client to AP; outdoors at up to 1,800 feet (600 meters) between antennas. 5 GHz systems: indoors at up to 900 feet (275 meters) from the client to AP; outdoors at up to 5,400 feet (1,800 meters) between antennas.
Total network devices	Up to 255 client devices may be associated with an AP, with 128 in simultaneous operation.
Voice connections	*Voice over Internet Protocol* (VoIP) (future)
Data security	Authentication: shared key or open key. Encryption: WEP at 64 bits or 128 bits for 2.4 GHz systems plus 152 bits for 5 GHz systems.
Modulation	2.4 GHz systems: DSSS. 5 GHz systems: OFDM.
48-bit network ID	Enables the concurrent operation of multiple co-located networks.

General Packet Radio Service (GPRS)

Analog cellular mobile phone systems are considered *first-generation* (1G) technology, whereas digital cellular mobile phone systems are referred to as *second-generation* (2G) technology. The next generation of cellular services offers a broadband data capability and is known simply as *3G*, for *third-generation* technology. When fully implemented, 3G technologies will make it possible for service providers to offer a variety of mobile services ranging from messaging to speech, data and video communications, Internet and intranet access, and high bit rate communication up to 2 Mbps.

Many carriers have already taken an interim step to 3G, referred to as *2.5G*, which uses IP to provide fast access to data networks via GPRS technology. Compared to *Circuit-Switched Data* (CSD), which operates at up to 14.4 Kbps, and *High-Speed Circuit-Switched Data* (HSCSD), which operates at up to 43.2 Kbps, GPRS utilizes packet-switching technology to transmit short bursts of data over an IP-based network to deliver speeds of up to 144 Kbps over an always-on wireless connection.

True 3G networks based on *Enhanced Data Rates for GSM Evolution* (EDGE) technology deliver data at speeds of up to 384 Kbps. Carriers in the United States have been moving toward 3G for several years by overlaying various technologies onto their existing networks to enhance their data-handling capabilities.

For carriers with TDMA-based networks, the first step to offering true 3G services is to deploy *Global Systems for Mobile communications* (GSM) and then GPRS. The new GSM/GPRS networks do not replace existing TDMA networks; carriers will continue supporting these networks long into the future to service their voice customers. Eventually, all TDMA customers will be migrated to GSM/GPRS. Once the GSM/GPRS overlay is in place in a market, the carriers can upgrade their networks with EDGE-compliant software to boost data transmission rates to as much as 384 Kbps and begin the availability of true 3G services.

Carriers whose wireless networks are based on *Code Division Multiple Access* (CDMA) will take a different technology path to 3G, going through CDMA2000, before eventually arriving at *Wideband CDMA* (W-CDMA). Both EDGE and W-CDMA offer a migration path to the global standard *Universal Mobile Telecommunications System* (UMTS).

Coverage for 2.5/3G services is still ramping up, despite the impressive figures thrown out by individual carriers. The next step is for service providers to engage in more roaming arrangements, which is a way to save costs, reduce time to market, and add value to attract more customers.

The data speed of 2.5/3G services is determined by many factors, including the equipment and software in the wireless network, the distance of the user from the nearest base station, and how fast the user may be moving. The claimed speed of the service is rarely, if ever, achieved in the real-world operating environment.

The pricing plans and price points differ by carrier, from a simple add-on to the existing digital voice plans for a basic data service to tiered pricing plans based on actual data usage. Depending on the applications, users can opt for 2.5/3G cell phones with multimedia messaging capabilities. Alternatively, users with heavy messaging and file transfer requirements may opt for PC Cards for notebooks and PDAs.

Some service providers intend to support WiFi as well as GPRS, viewing them as complementary. The existing GPRS and upcoming EDGE networks provide wide area coverage for applications where customers want brief access to applications such as their e-mail and calendar, whereas WiFi networks will be available in convenient locations where customers are likely to spend time accessing larger data files and browsing the Web. Service providers will offer seamless access to 3G and WiFi networks via one PC Card (see Figure 1-7) and bill customers for both services with one invoice.

Like WiFi, one of the characteristics of 3G wireless technology is always-on high-speed mobile Internet connection. Lucent Technologies has demonstrated the successful seamless hand-off of a wireless data call from a WiFi to a GSM-based 3G network and WiFi to a CDMA2000-based 3G network, enabling mobile laptop users to browse the Internet while roaming between the two network types with no interruption in the session. The capabilities rely on the Mobile IP standard from the *Internet Engineering Task Force* (IETF). Mobile IP supports intertechnology handoffs between WiFi and 3G technologies. As a future service, this will enhance mobile workers' wireless experience by giving them continuous access to the information they need using the fastest technology available in any given location, greatly extending the wireless coverage area.

Figure 1-7 Nokia offers the D211 GSM card, providing GPRS and WLAN connectivity in one. Another version of the product, the D311, is designed for the North American market. Drivers are available for PocketPC, Windows, and Linux.

Fixed Wireless Access

Fixed wireless access technology provides a wireless link to the PSTN using spectrum licensed by the FCC. It constitutes an alternative to traditional wire-based local telephone service. Since calls and other information (for example, data and images) are transmitted through the air rather than through conventional cables and wires, the cost of providing and maintaining telephone poles and cables is avoided. Unlike cellular technologies, which provide services to mobile users, fixed wireless services require a rooftop antenna to an office building or home, which is lined up with a service provider's hub antenna. Although WiFi is set up in a similar manner, it uses unlicensed spectrum and does not provide connectivity with telephone networks.

Fixed wireless access systems come in two varieties: narrowband and broadband. A narrowband fixed wireless access service can provide bandwidth up to 128 Kbps, which can support one voice conversation and a data session such as Internet access or fax transmission. A broadband fixed wireless access service can provide bandwidth in the multimegabit-per-second range, which is enough to support telephone calls, television programming, and broadband Internet access.

A narrowband fixed wireless service requires a wireless access unit, which is installed on the exterior of a home or business (see Figure 1-8) to enable customers to originate and receive calls without changing their

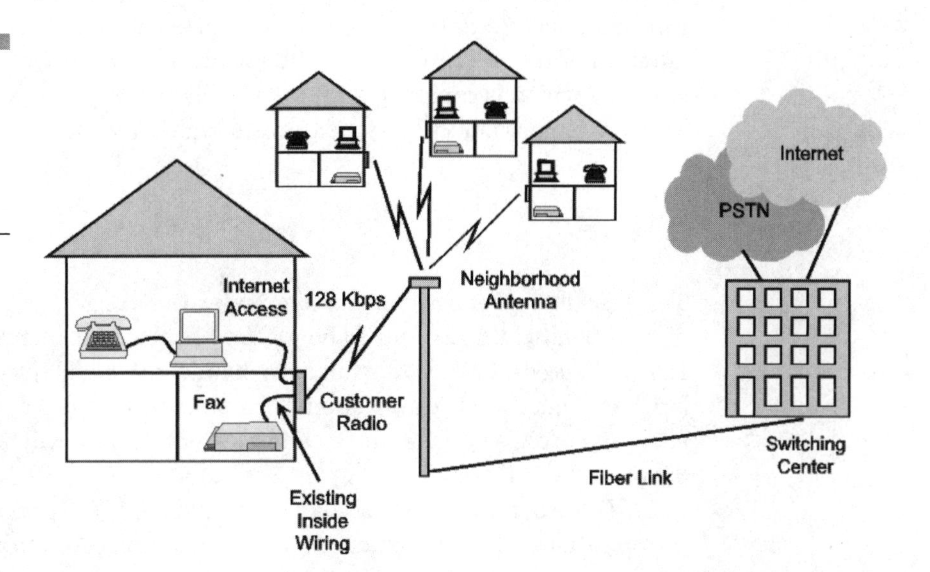

Figure 1-8 A simple fixed wireless configuration for a narrowband access service

existing analog telephones. This transceiver is positioned to provide an unobstructed view to the nearest base station receiver. Voice and data calls are transmitted from the transceiver at the customer's location to the base station equipment, which relays the call through carrier's existing network facilities to the appropriate destination. No investment in special phones or facsimile machines is required; customers use all their existing equipment.

Narrowband fixed wireless systems use the licensed 3.5 GHz radio band with 100 MHz spacing between the uplink and downlink frequencies. Subscribers receive network access over a radio link within a range of 200 meters (600 feet) to 40 kilometers (25 miles) of the carrier's hub antenna. About 2,000 subscribers can be supported per cell site.

Broadband fixed wireless access systems are based on microwave technology. *Multichannel Multipoint Distribution Service* (MMDS) operates in the licensed 2 to 3 GHz frequency range, whereas *Local Multipoint Distribution Service* (LMDS) operates in the licensed 27 to 31 GHz frequency range. Both services are used by *Competitive Local Exchange Carriers* (CLECs) primarily to offer broadband Internet access, but they are also capable of supporting voice. These technologies are used to bring voice and data traffic to the fiber-optic networks of *Interexchange Carriers* (IXCs) and nationwide CLECs, bypassing the local loops of the *Incumbent Local Exchange Carriers* (ILECs).

Even if a business subscribes to MMDS or LMDS, it can still use WiFi within a building or campus environment. Once the traffic moves from the WiFi link to the wired LAN via an AP, it can go out to an MMDS/LMDS link via a hub or switch. The customer's hub or switch would be connected to a *network interface unit* (NIU), which is cabled to a rooftop MMDS/LMDS antenna. That antenna sends the traffic to the service provider's hub antenna, which is cabled to an *Asynchronous Transfer Mode* (ATM) switch. That switch provides high-speed access to the Internet (see Figure 1-9).

LMDS

This broadband service enables communications providers to offer a variety of high-bandwidth services to homes and businesses, including broadband Internet access. LMDS offers greater bandwidth capabilities than MMDS, but has a maximum range of only 7.5 miles from the carrier's hub to the customer premises. This range can be extended, however, through the use of optical fiber links.

LMDS provides enormous bandwidth: enough to support 16,000 voice conversations plus 200 channels of television programming. CLECs can

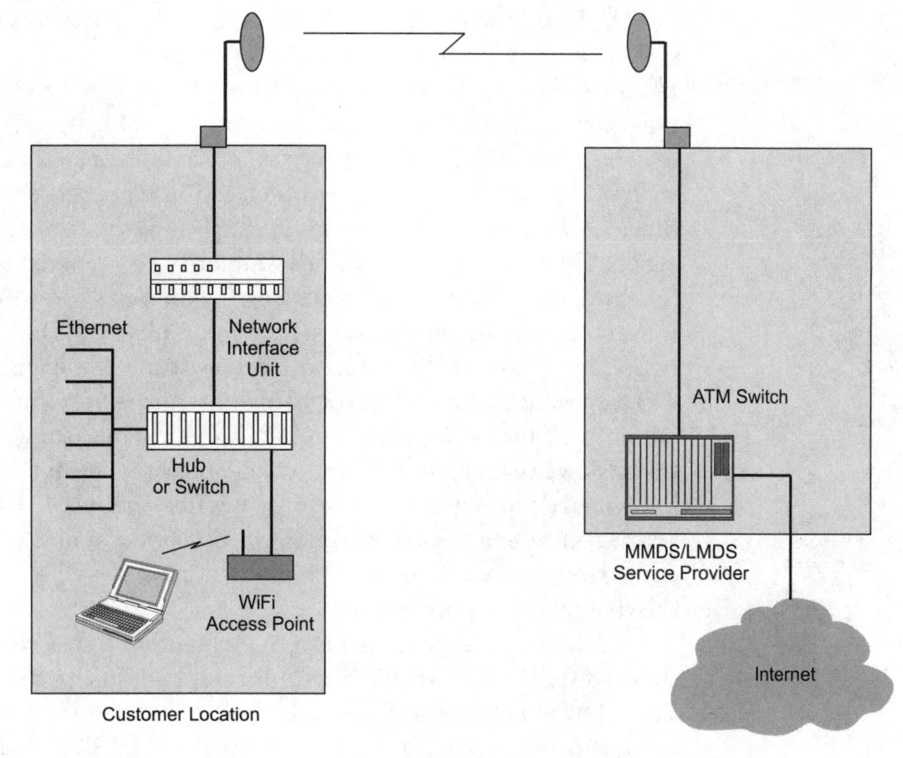

Figure 1-9 Data from a WiFi connection can reach the Internet through other wireless connections, such as MMDS or LMDS.

deploy LMDS to completely bypass the local loop, eliminating access charges and avoiding service-provisioning delays. Because the service entails setting up equipment between the provider's hub location and customer buildings for the microwave link, LMDS costs far less to deploy than installing new fiber. This enables CLECs to economically bring customer traffic onto their existing metropolitan fiber networks and, from there, to a national backbone network.

The strategy among many CLECs is to offer LMDS to owners of multitenant office buildings and then install cable to each tenant who subscribes to the service. The cabling goes to an on-premises switch, which is run to the antenna on the building's roof. That antenna is aimed at the service provider's antenna at its hub location. The line-of-sight wireless link between the two antennas offers a broadband pipe for multiple voice, data, and video applications. Subscribers can use LMDS for a variety of high-bandwidth applications, including television broadcast, videoconferencing, LAN interconnection, broadband Internet access, and telemedicine.

LMDS operation requires a clear line of sight between the carrier's hub station antenna and the antenna at each customer location. However, LMDS is also capable of operating without having a direct line of sight with the receiver. This feature, which is highly desirable in built-up urban areas, may be achieved by bouncing signals off buildings so that they get around obstructions. At the receiving location, the data packets arriving at different times are held in queue for resequencing before they are passed to the application. This scheme does not work well for voice, however, because the delay resulting from queuing and resequencing disrupts two-way conversation.

A roof-mounted multisectored antenna is placed at the carrier's hub location. Each sector of the antenna receives/transmits signals between itself and a specific customer location. This antenna is very small, often measuring only 12 inches in diameter. The hub antenna brings the multiplexed traffic down to an indoor switch, which processes the data into 53-byte ATM cells for transmission over the carrier's fiber network. These individually addressed cells are converted back to their native format before going off the carrier's network to their proper destinations—the Internet, PSTN, or the customer's remote location.

Each customer's location has a rooftop antenna that sends/receives multiplexed traffic. This traffic passes through an indoor NIU, which provides the gateway between the RF components and the in-building equipment, such as a LAN hub, *private branch exchange* (PBX), or videoconferencing system. The NIU includes an up/down converter that changes the frequency of the microwave signals to a lower *intermediate frequency* (IF) that the electronics in the office equipment can more easily (and inexpensively) manipulate.

A potential problem for LMDS users is that the signals can be disrupted by heavy rainfall and dense fog—even foliage can block a signal. In metropolitan areas where new construction is a fact of life, a line-of-sight transmission path can disappear virtually overnight. For these reasons, many IT executives are leery of trusting mission-critical applications to this wireless technology. Service providers downplay this situation by claiming that LMDS is just one local access option and that fiber links are the way to go for mission-critical applications. In fact, some LMDS providers offer fiber as a backup in case the microwave links experience interference.

There is controversy in the industry about the economics of the point-to-multipoint architecture of LMDS. Some experts claim that the business model of going after low-usage customers is fundamentally flawed and will never justify the service provider's cost of equipment, installation, and provisioning. With an overabundance of fiber in the ground and metropolitan area Gigabit Ethernet services coming online at a competitive price, the time for LMDS may have come and gone. In addition, newer wireless technologies like free-air laser hold a significant speed advantage over LMDS, as do submillimeter transmission in the 60 and 95 GHz bands.

Fiber optics is the primary transmission medium for broadband connectivity today. However, of the estimated 4.6 million commercial buildings in the United States, 99 percent are not served by fiber. Businesses are at a competitive disadvantage in today's information-intensive world unless they have access to broadband access services, including high-speed Internet access. These businesses, including many data-intensive high-technology companies, can be adequately served with LMDS. Despite the financial problems of LMDS providers, the technology has the potential to become a significant portion of the global access market, which will include a mix of many technologies, including DSL, cable modems, broadband satellite, and fiber-optic systems.

MMDS

This microwave technology traces its origins to 1972 when it was introduced to provide an analog service called *Multipoint Distribution Service* (MDS). For many years, MMDS was used for the one-way broadcast of television programming, but in early 1999, the FCC opened up this spectrum to allow two-way transmissions, making it useful for delivering telecommunication services, including high-speed Internet access to homes and businesses.

This technology, which has now been updated to digital, operates in the 2 to 3 GHz range, enabling large amounts of data to be carried over the air from the operator's antenna towers to small receiving dishes installed at each customer location. The useful signal range of MMDS is about 30 miles, which beats LMDS at 7.5 miles and DSL at 18,000 feet. Furthermore, MMDS is easier and less costly to install than cable service.

With MMDS, a complete package of television programs can be transmitted to homes and businesses. Because MMDS operates within the frequency range of 2 to 3 GHz, which is much lower than LMDS at 28 to 31 GHz, it can support only up to 24 stations. However, operating at a lower frequency range means that the signals are not as susceptible to interference as those using LMDS technology.

Most of the time the operator receives television programming via a satellite downlink. Large satellite antennas installed at the head end collect these signals and feed them into encoders that compress and encrypt the programming. The encoded video and audio signals are modulated via *amplitude modulation* (AM) and *frequency modulation* (FM), respectively, to an IF signal. These IF signals are up-converted to MMDS frequencies, and then amplified and combined for delivery to a coax cable, which is connected to the transmitting antenna. The antenna can have an omnidirectional or sectional pattern.

The small antennas at each subscriber location receive the signals and pass them via a cable to a set-top box connected to the television. If the service also supports high-speed Internet access, a cable also goes to a special modem connected to the subscriber's PC. MMDS sends data as fast as 10 Mbps downstream (toward the computer). Typically, service providers offer downstream rates of 512 Kbps to 2.0 Mbps, with burst rates up to 5 Mbps whenever spare bandwidth becomes available.

Originally, there was a line-of-sight limitation with MMDS technology. But this has been overcome with a complementary technology called *vector orthogonal frequency division multiplexing* (VOFDM). Because MMDS does not require an unobstructed line of sight between antennas, signals bouncing off objects en route to their destination require a mechanism for being reassembled in their proper order at the receiving site. VOFDM handles this function by leveraging multipath signals, which normally degrade transmissions. It does this by combining multiple signals at the receiving end to enhance or recreate the transmitted signals. This increases the overall wireless system performance, link quality, and availability. It also increases service providers' market coverage through non-line-of-sight transmission.

MMDS equipment can be categorized into two types based on the duplexing technology used: *frequency division duplexing* (FDD) or TDD. Systems based FDD are good solutions for voice and bidirectional data because forward and reverse use separate and equally large frequency bands. However, the fixed nature of this scheme limits the overall efficiency when used for Internet access. This is because Internet traffic tends to be bursty and asymmetrical. Instead of preassigning bandwidth with FDD, Internet traffic is best supported by a more flexible bandwidth allocation scheme.

This is where TDD comes in; it is more efficient because each radio channel is divided into multiple time slots through TDMA technology, which enables multiple channels to be supported. Because TDD has flexible time slot allocations, it is better suited for data delivery—specifically, Internet traffic. TDD enables service providers to vary uplink and downlink ratios as they add customers and services. Many more users can be supported by the allocation of bandwidth on a nonpredefined basis.

MMDS is being used to fill the gaps in market segments where cable modems and DSL cannot be deployed because of distance limitations and cost concerns. Like these technologies, MMDS provides data services and enhanced video services such as video on demand as well as Internet access. MMDS can be another access method to complement a carrier's existing cable and DSL infrastructure, or it can be used alone for direct competition. With VOFDM technology, MMDS is becoming a workable option that can be deployed cost effectively to reach urban businesses that do have line-of-sight access, and in suburban and rural markets for small businesses and telecommuters.

Fixed wireless access technology originated out of the need to contain carriers' operating costs in rural areas, where pole and cable installation and maintenance are more expensive than in urban and suburban areas. However, wireless access technology can also be used in urban areas to bypass the LEC for long-distance calls. Since the IXC or CLEC avoids having to pay the ILEC's local loop interconnection charges, the savings can be passed back to the customer. This arrangement is also referred to as a *wireless local loop* (WLL).

Laser Transmission

A relatively new category of wireless communication uses laser, sometimes called *free-space optics*, operating in the near-infrared region of the light spectrum. Utilizing coherent laser light, these wireless line-of-sight links are used to link buildings in campus environments and urban areas where the installation of cable is impractical and the performance of leased lines is too slow.

Laser links between sites can be operated at the full LAN channel speed. In addition, unlike microwave transmission, laser transmission does not require an FCC license, and data traveling by laser beam cannot be intercepted. Via an AP to the wired LAN, WiFi traffic within a building can go out a hub or switch connected to the laser system. From there, the data is beamed to another building's laser system connected to its LAN.

The lasers at each location are aligned with a simple bar graph and tone lock procedure. Fiber-optic repeaters are used to connect the LANs to the laser units. Alternatively, a bridge equipped with a fiber-optic-to-AUI transceiver can be used (see Figure 1-10). Connections to and from the laser are made using standard fiber-optic cable, protecting data from sources of RF and EMI. Monitors can be attached to the laser units to provide operational status, such as signal strength, and to implement local and remote loopback diagnostics.

Figure 1-10 Terabeam Magna, a free-space optics system from TeraBeam Corporation

The reason why laser products are not used very often for business applications is because transmission is diminished by atmospheric conditions that produce effects such as absorption, scattering, and shimmer. All three can reduce the amount of light energy that is picked up by the receiver and corrupt the data being sent.

Absorption refers to the capability of various frequencies to pass through the air. Absorption is determined largely by the water vapor and carbon dioxide content of the air along the transmission path, which, in turn, depends on humidity and altitude. The gases that form in the atmosphere have many resonant bands, which enable specific frequencies of light to pass. These transmission windows occur at various wavelengths, such as the visible light range. Another window occurs at the near-infrared wavelength of approximately 820 nm. Laser products tuned to this window are not greatly affected by absorption.

Scattering has a much greater effect on laser transmission than absorption. The atmospheric scattering of light is a function of its wavelength and the number and size of scattering particles in the air. The optical visibility along the transmission path is directly related to the number and size of these particles. Fog and smog are the main conditions that tend to limit visibility for optical-infrared transmission followed by snow and rain.

Shimmer is caused by localized differences in the air's index of refraction. This is caused by a combination of factors, including the time of day (daytime heat), terrain, cloud cover, wind, and the height of the optical path above the source of shimmer. These conditions cause fluctuations in the received signal level by directing some of the light out of its intended path. Beam fluctuations may degrade system performance by producing short-term signal amplitudes, which approach threshold values. Signal fades below these threshold values result in error bursts.

Vendors have taken steps to mitigate the effects of absorption, scattering, and shimmer. For example, techniques such as FM in the transmitter and an *automatic gain control* (AGC) in the receiver can minimize the effects of shimmer. Also, selecting an optical path several meters above heat sources can greatly reduce the effects of shimmer. However, all of these distorting conditions can vary greatly within a short time span or persist for long periods, requiring on-site expertise to constantly fine-tune the system.

Many businesses simply cannot risk frequent or extended periods of downtime while the necessary compensating adjustments are being made. As if all this was not enough, one must contend with other potential problems, such as thermal window coatings and the laser beam's angle of incidence, both of which can disrupt transmission. These problems are being overcome with newer lasers that operate in the 1,550 nm wavelength. A 1,550 nm delivery system is powerful enough to go through windows, can deliver signals under the fog blanket, and is safe enough that it does not blind the casual viewer who happens to look into the beam. Up to 1 Gbps of bandwidth is available with these systems—the equivalent bandwidth capacity of 660 T1 lines.

Laser also carries a distance limitation associated with laser. The link generally cannot exceed 1.5 km, and 1 km is preferred. With 1,550 nm systems, the practical distance of the link is only 500 meters. Despite its limitations, laser (or free-space optics) can provide a valuable last link between the fiber network and the end user—including serving as a backup to more conventional methods such as fiber. Free-space optics, unlike other transmission technologies, is not tied to standards or standards development. Vendors simply attach their equipment into existing fiber-based networks and then use any laser transmission methods they like. This encourages innovation, differentiation, and speed of deployment.

Conclusion

The emergence and proliferation of WiFi networks has become a phenomenal pull on broadband demand. In fact, the sudden success of WiFi has the FCC concerned that the airwaves may be getting too crowded, especially as *wireless Internet service providers* (WISPs) extend the range of WiFi to broaden their coverage areas. Amateur radio enthusiasts and some television stations, for example, claim that WiFi products are raising the level of interference on their transmissions. According to some interpretations of FCC regulations, since amateur radio and television stations are licensed, users of offending unlicensed WiFi gear must either eliminate the interference or shut down. Court cases have upheld this interpretation, which basically holds that licensed operators have preference over unlicensed operators.

As more devices come online to take advantage of WiFi, the FCC will come under more pressure to adopt measures that enable devices to peacefully coexist in the shared spectrum, such as reducing the power of 2.4 GHz devices to limit RF emissions or refining the definition of spread spectrum to make it less interfering. In the past, the FCC has even migrated operators from one spectrum band to another if it was deemed to be in the public interest. Judging by the impact it is already having on Internet users, it is more likely that the FCC will give preference to WiFi as it looks for ways to accommodate all users.

CHAPTER 2

Wireless Network Components

Wireless Fidelity (WiFi) technology is becoming a widely accepted part of the enterprise technology portfolio. Removing the wires necessary for a *local area network* (LAN) is a natural evolution in the business environment. It lets businesses create a new, fully configured network or extend an existing wired network in minutes without having to run cables to desktops. Wireless technology offers all the functionality of current LANs—information exchange and high-speed connection—but without the constraints of wires. This enables users to roam throughout the office while maintaining a wireless connection to the company's network and the Internet. Several components go into building a wireless enterprise network.

A basic configuration generally consists of a computer, printer, handheld, or similar independent device that is equipped with a low-power WiFi (802.11b) or WiFi5 (802.11a) radio. WiFi radios come in many types of platforms including *access points* (APs), bridges, routers, and gateways, as well as *Peripheral Component Interconnection* (PCI), Mini PCI, PC Card, and *universal serial bus* (USB) devices. More recently, WiFi radios have become available in the Compact Flash form factor. Some of these devices are even available in dual-mode versions, supporting WiFi at both the 2.4 and 5 GHz frequency bands, supporting WiFi and Bluetooth simultaneously, or supporting WiFi and 2.5G/3G cellular data services such as *General Packet Radio Service* (GPRS). For difficult indoor environments, extender antennas are available that can be strategically placed to increase range by 15 to 30 percent. For campus environments, outdoor antennas are available that can ensure uninterrupted WiFi connections even as employees move between buildings.

PC Cards for Notebooks

PC Cards insert into a Type II slot of notebook computers and handheld devices. For wireless connectivity, these cards contain an integral WiFi radio and antenna. Many of these cards are priced very inexpensively for consumers and do not provide much in the way of management features. However, enterprise users may need cards that support extensive management features implemented from an administrator's console, and even though they are priced higher than consumer products, they are well worth it. For example, cards that come with a site survey tool scan for and display APs by their domain access name and *Media Access Control* (MAC) address ID. They show the *Wired Equivalent Privacy* (WEP) status, signal quality, and the estimated distance to help place APs for optimum performance.

Another management feature enables the user to create a list that prioritizes the available APs for association and, when used by a network administrator, facilitates the planning of wireless links and network traffic balancing. The user can also manually select an AP to associate with. This provides a quick alternative for users who want to link up fast without going through all the available prioritized APs.

For telecommuters, a network profile manager is available that enables network configurations for different network environments to be saved—one for the home, one for the office, and possibly others for various public APs. This saves time and minimizes the network configuration effort as the user moves through different environments throughout the day. Table 2-1 summarizes the typical performance users can expect from a WiFi PC Card in indoor and outdoor environments.

Some WiFi vendors also offer dual-band PC Cards that support both 802.11a and 802.11b standards, enabling users to connect to any wireless network as they move through different locations or run different applications. To use the card effectively, the management software that comes with the dual-band card enables users to create profiles that reflect specific *wireless LAN* (WLAN) settings for different locations and provides a station listing of all available APs.

Table 2-1

Typical performance of WiFi PC Cards in indoor and outdoor environments

Indoor Operating Range	Distance	Transmission Speed
	Up to 100 feet (30 meters)	11 Mbps
	Up to 165 feet (50 meters)	5.5 Mbps
	Up to 230 feet (70 meters)	2 Mbps
	Up to 300 feet (91 meters)	1 Mbps
Outdoor Operating Range	**Distance**	**Transmission Speed**
	Up to 500 feet (152 meters)	11 Mbps
	Up to 885 feet (270 meters)	5.5 Mbps
	Up to 1,300 feet (396 meters)	2 Mbps
	Up to 1,500 feet (457 meters)	1 Mbps

PCI Cards for Desktops

PCI Cards for installation into tower and desktop computers come in several versions. D-Link, for example, offers a PCI adapter with a detachable antenna that can transmit data at rates of 11, 5.5, 2, and 1 Mbps (see Figure 2-1). It also supports 64- and 128-bit WEP encryption for network security. The antenna has an effective range of up to 230 feet indoors and up to 984 feet outdoors. The detachable antenna design provides the flexibility to connect a higher-gain antenna for better transmission quality over a wider coverage area or in a *radio frequency* (RF) hostile environment.

Another kind of PCI adapter is essentially a simple holder for a standard PC Card normally used with a notebook computer (see Figure 2-2). The PC Card actually has a built-in WiFi radio and antenna, whereas the PCI adapter acts as a PC Card reader. When the tower or desktop computer is not in use, the PC Card can be easily removed from the PCI adapter and inserted into the notebook's slot, giving telecommuters a convenient way to take advantage of WiFi with minimal investment in equipment. Because the WiFi antenna is built into the PC Card, however, users do not have the option of attaching an external antenna to improve the signal and increase its range.

Figure 2-1 This PCI adapter from D-Link features a detachable antenna that can be replaced with a higher-powered antenna to increase the range and signal quality of a WiFi link.

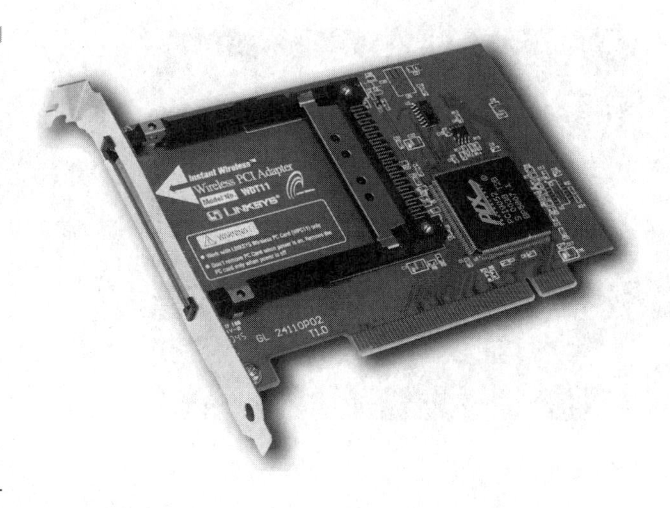

Figure 2-2 The Instant Wireless PCI Adapter from Linksys acts as a PC Card reader, giving users the freedom to work from either a desktop or notebook computer with minimal investment in equipment.

Mini PCI Cards

Early adapters of WiFi often had a difficult time manually configuring their notebook computers for wireless networking. Making changes to their notebook often resulted in an inoperable network connection. Dell, Hewlett-Packard, IBM, Sony, and Toshiba are among the growing number of computer companies that offer integrated WiFi capabilities in their notebooks, putting an end to manual configurations. The WiFi antennas are embedded directly into the lid of the notebook for optimal signal strength, eliminating the need for a PC Card.

The WiFi radio is contained on a very small card called the Mini PCI, which is functionally identical to standard desktop computer PCI Cards but can be slotted on the computer's main board (see Figure 2-3). Many laptop manufacturers have adopted the Mini PCI form factor to add integral support for wireless networking and modems, enabling PC Card slots to remain free for other options.

The Mini PCI Card offers several advantages over current proprietary and vendor-specific integrated communications devices, which are not designed to be replaced when they fail or upgraded when technology changes. In contrast, depending on the particular system design, a service technician can replace a standard Mini PCI Card if it fails or needs an upgrade rather than replacing the entire system board, which entails a higher cost to the customer.

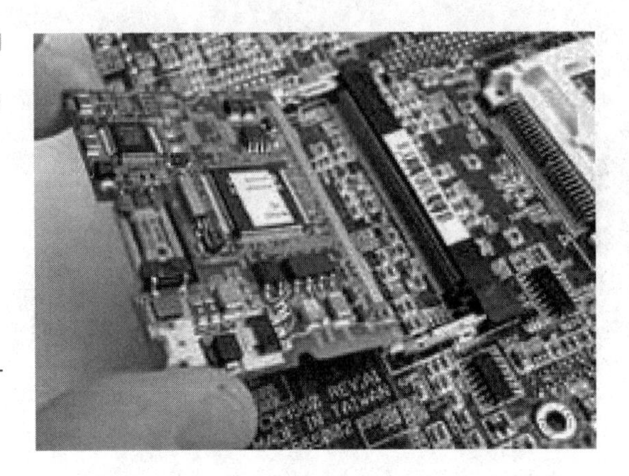

Figure 2-3
Advantech offers a 56 Kbps modem in the Mini PCI form factor. Other vendors offer an integrated WiFi radio and antenna in this form factor.

Mini PCI Cards can support other wireless technologies, including Bluetooth and GPRS, as well as wire technologies such as G.lite, a variation of *Asymmetrical Digital Subscriber Line* (ADSL), and IEEE 1394, also known as FireWire. Some Mini PCI Cards also support multiple communications technologies. For example, one Mini PCI Card could provide both WiFi and Bluetooth connectivity. WLAN chip-set provider Intersil and Bluetooth radio maker Silicon Wave have devised a reference design called Blue802, which allows for the near simultaneous operation of the Bluetooth and WiFi wireless protocols.

A notebook equipped with a Blue802 Mini PCI Card could run a video stream to a WLAN AP while sending data to a printer via Bluetooth. This also enables PC-based Bluetooth applications such as mouse, keyboard, printing, file transfer, and portable device synchronization to run at the same time the user's PC is connected to a network over WiFi.

Bluetooth and WiFi operate in the same 2.4 GHz radio band, but because Bluetooth employs frequency hopping, it jumps all over the band, slowing down and even terminating WiFi connections. The Blue802 technology overcomes this obstacle by using a time-slicing technique in which the two protocols are not actually running simultaneously, but switch back and forth so fast that the connection seems simultaneous.

USB Adapters

A USB adapter provides wireless connectivity through a computer's USB port (see Figure 2-4). This is the simplest way to add wireless connectivity

Figure 2-4 ActionTec's 11 Mbps USB Wireless Adapter enables users to connect any computer with an available USB port into a wireless network.

to desktop computers and any peripheral equipment with a USB port. Legacy equipment, printers, and other devices that lack PCI slots into which an enabling WiFi radio could be added may use a USB-based WiFi radio instead. The adapter is powered by the computer's USB port, eliminating the need for an external power supply. Additionally, some USB adapters come with cable so users can place the unit with its incorporated antenna up on top of the desk to improve signal range.

After it is installed, the USB adapter can be easily configured through a control panel that is accessible via an icon in the Windows task bar. The control panel enables users to adjust settings, check signal strength and link quality, perform a site survey, monitor statistics, and adjust advanced features. The adapter supports 64- and 128-bit WEP data encryption, which is also configurable through the same control panel.

Compact Flash Cards

Mobile professionals with *personal digital assistants* (PDAs) and other handheld devices are also among the users who require fast, cable-free communications to network applications throughout the enterprise. The Compact Flash form factor is designed to give small devices additional functionality, including WiFi connectivity. This is the same form factor that provides memory for digital cameras, except that the card contains an integrated WiFi radio and antenna.

Symbol Technologies, for example, is among the vendors offering a Compact Flash card for WiFi called the Wireless Networker, which is suitable for use in PocketPC/Windows CE PDAs, as well as mobile bar-code-scanning products (see Figure 2-5). A signal range of up to 300 feet (91 meters) is supported in standard office environments. The card features automatic rate

Figure 2-5 The Wireless Networker Compact Flash card (left) from Symbol Technologies provides PDAs and other portable devices with WiFi connectivity (right).

scaling at 11, 5.5, 2, and 1 Mbps for maximum range. Data security is achieved with 64- and 128-bit WEP data encryption.

The Compact Flash card extends battery life by allowing the WiFi radio to idle when it is not active. In addition, to prevent accidental battery drain, a suspend feature enables PocketPCs to automatically turn off the Wireless Networker based on screen activity when data is not being transmitted.

Secure Digital (SD) Cards

Support for WiFi has been incorporated into *Secure Digital* (SD) cards for handheld devices. SD is the smallest of the major removable card formats —not much bigger than a postage stamp. SD cards and the SD slot were originally intended as mechanisms for removable storage, but with the ratification of the *secure digital input/output* (SDIO) specification by the SD Association in 2002, the way has been cleared for incorporating new features into handheld devices such as wireless networking (for example, Bluetooth and WiFi) and global positioning systems.

Wireless cards for the SD slot essentially make it much easier for manufacturers to provide 802.11b connectivity for PDAs and other devices such as cell phones without the added bulk and cost of adapter sleeves. Users can also share Internet access and network resources by trading the card back and forth between their own or another user's devices. At this writing, PDAs such as Hewlett-Packard's Compaq iPAQ and several models from Palm come with an SD slot.

As more uses for the SD form factor become available, the popularity of SD should increase to the point where most handheld devices will come equipped with the card slot. The primary benefits of SD cards are speed, *input/output* (I/O) capabilities, security, and interoperability. These features make it among the most secure and flexible media to expand the capabilities of handheld devices and facilitate interoperability with other devices. At 10 Mbps, it is a very fast interface for both memory and I/O.

PDA Modules

Some PDAs that do not have slots for Compact Flash or SD cards offer WiFi connectivity through the use of modules that attach to the device. For the Handspring Visor handheld computer, for example, the Xircom division of Intel offers the SpringPort Wireless LAN Module. The module plugs into the Springboard expansion slot on the back of the Handspring Visor (see Figure 2-6). The product supports up to three wireless network profiles, which lets the user define different settings for various APs. The module adds bulk to the Visor, but this is because it contains a lithium-ion battery, which supplies power to the module without draining the PDA's batteries.

The trend of embedding WiFi into notebooks is coming to PDAs as well. Toshiba's PocketPC e740 is one PDA that sports integrated WiFi connectivity, enabling users to do things such as surf the Internet, check e-mail, and interact with the local network without wires or cables (see Figure 2-7).

Figure 2-6 The SpringPort Wireless LAN Module attaches to Handspring Visor handhelds.

Figure 2-7 Toshiba's PocketPC e740 comes with a built-in WiFi radio, enabling its Compact Flash card slot to be used for other purposes.

The e740 also offers a built-in Compact Flash expansion slot, but since it offers built-in WiFi, the free slot can be used for increasing the device's functionality, storage, or connectivity options.

Soft WiFi

Under the concept of Soft WiFi, which originated with Microsoft, a number of tasks routinely given to the radio, the most expensive of several WiFi components in any given device, are reassigned to either the modem that lets notebooks access a WiFi network or to the notebook itself. Rather than seeing just Ethernet header information, such as the destination and source addresses, Soft WiFi exposes the raw WiFi data stream to Windows. This enables a Windows PC to act as an AP. By offloading some of the 802.11 processing from the AP to Windows, Microsoft offers less expensive hardware for consumers.

Microsoft had previously applied a similar concept to modems called *WinModem*. Offloading processing from the modem adapter to the PC itself enabled manufacturers to make cheaper adapters. But such strategies also tend to increase users' dependence on Windows.

Intel has articulated a similar plan called *Soft AP*. Although chip makers believe today's PCs can handle the workload of a hardware AP, they are not so sure WiFi won't slow to a crawl under Windows. To make up for Windows' lack of real-time processing, Intel is developing a way to split an AP between Windows and a chip for 802.11 cards to handle real-time events.

Both Intel and Microsoft are targeting consumers rather than businesses. Intel has stated that its Soft AP could cut the cost of consumer APs in half to just $100. An enterprise AP usually includes central management features, which accounts for its higher cost.

Access Points (APs)

An AP provides the connection between one or more wireless client devices and a wired LAN. The AP is usually connected to the wired LAN via a *Category 5* (CAT 5) cable connection to a hub or switch. Client devices communicate with the AP over the wireless link, giving them access to all other devices through the hub or switch, including a router on the other side of the hub, which provides access to the Internet, *virtual private networks* (VPNs), and corporate intranets (see Figure 2-8).

An AP that adheres to the 802.11b standard for operation over the unlicensed 2.4 GHz band supports a wireless link with a data transfer speed of up to 11 Mbps, whereas an AP that adheres to the 802.11a standard for operation over the unlicensed 5 GHz band supports a wireless link with a data transfer speed of up to 54 Mbps. APs include a number of functions and features, including the following:

- Radio power control for flexibility and easy networking setup.
- Dynamic rate scaling, Mobile IP functionality, and advanced transmit/receive technology to enable multiple APs to serve users on the move.
- Built-in bridging and repeating features to connect buildings miles apart. The use of specialty antennas increases range. The AP can support simultaneous bridging and client connections.
- WEP, which helps protect data in transit over the wireless link between the client device and the AP via 64-, 128-, or 256-bit encryption.
- *Access control list* (ACL) and VPN compatibility, which help guard the network from intruders.

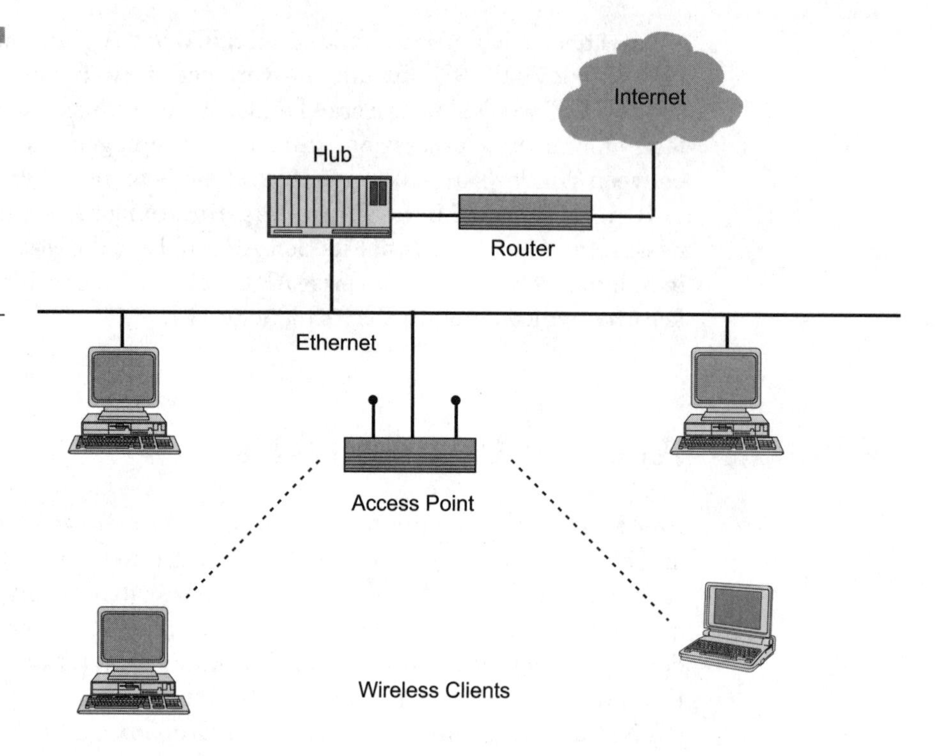

Figure 2-8 A simple configuration showing the relationship of the AP to the wired and wireless segments of the network

- Statistics on the quality of the wireless link (see Figure 2-9).
- Configurable using the embedded web browser.

APs can even be connected to each other, or be connected via specialized wireless bridges, to create daisy-chained networks. This extends the reach of wireless links to ensure campus coverage, for example, and enables users to roam over greater distances without breaking the connection. APs can also be connected to outdoor routers for full mesh connectivity between buildings.

Consumer Versus Commercial

APs are available in consumer and commercial versions. The latter generally cost more because of their extensive management capabilities, troubleshooting features, and power options. They usually support more security features as well. The emerging 802.1x standard for user authentication, for example, enables APs to pass login information to a *Remote Authentication Dial-In User Service* (RADIUS) server rather than handling

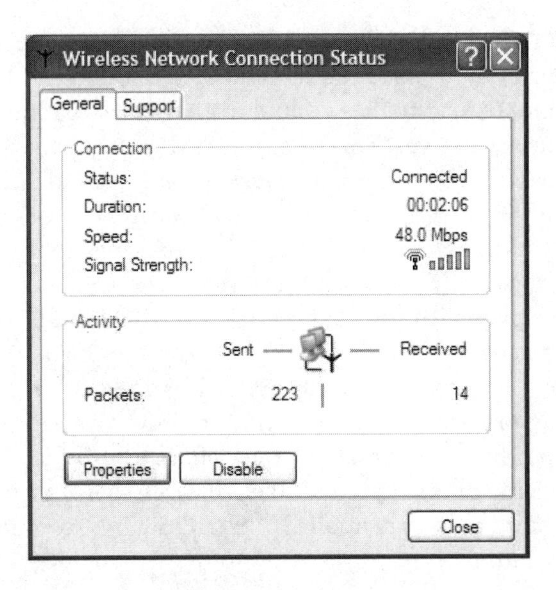

Figure 2-9 The 5 GHz DWL-5000 AP from D-Link Systems, Inc. keeps the client device notified of the status of the wireless link. In this case, the signal is at near maximum strength, providing a data transfer rate of 48 Mbps.

it directly or through a local system connected to the AP. RADIUS is an example of a security management protocol that lets an administrator control which remote users connect to the corporate network and which resources they can access.

Another difference between consumer and commercial products is that the commercial products are usually more solidly built. Whereas consumer products have plastic cases, commercial equipment features a more rugged metal case and can run in a broader operating temperature range, making it better suited for installation in harsh environments such as factories and warehouses.

Although both consumer and commercial products are available in 2.4 and 5 GHz versions, some commercial units feature two card slots, enabling WiFi and WiFi5 clients to be supported from the same AP. The choice of a dual-band AP provides organizations with a migration path to the higher data transfer speeds available with 802.11a while continuing to support their existing investment in 802.11b infrastructure. Depending on the manufacturer, these dual-band APs are modular so they can be upgraded to support future 802.11 technologies as they become available, which further protects an organization's investment in wireless infrastructure.

Cisco's Aironet 1200 Series AP, for example, protects the existing investment in WiFi networks while delivering a seamless migration path to WiFi5 and emerging technologies such as IEEE 802.11g, which provides up to 54 Mbps over WiFi's 2.4 GHz frequency band. The modular design of the Cisco

Aironet 1200 Series AP allows for both single- and dual-radio configurations for operation in both the 2.4 and 5 GHz unlicensed bands plus field upgradeability to modify these configurations as user requirements develop and the industry evolves. This dual-radio AP costs about $1,500, whereas a single-radio AP for the consumer market costs around $200.

Consumer-level APs stress the ease of setup and use (see Figure 2-10). Many products are configured with default settings that enable the user to plug in the device and establish the wireless connection immediately. Later, the user can play with the configuration settings to improve performance and set up security.

Although APs adhere to the 802.11 standards, manufacturers can include some proprietary features that improve the data transfer speed of the wireless link. For example, D-Link offers a *turbo mode* that optionally increases the maximum speed of 802.11b wireless links from 11 to 22 Mbps. When this turbo feature is applied to 802.11a wireless links, data can burst from 54 to 72 Mbps.

Systems Management

Enterprise-level APs provide more management features, enabling LAN administrators to remotely set up and configure multiple APs and clients from a central location. For monitoring and managing an entire WLAN infrastructure consisting of hundreds or even thousands of APs, however, a

Figure 2-10 This 5 GHz Wireless AP from Linksys, which features an antenna with a range of up to 328 feet indoors, is an example of a consumer AP.

dedicated management system is usually required. Such systems automatically discover every AP on the network and provide real-time monitoring of an entire wireless network spread out over multiple facilities and subnets. These management systems support the *Simple Network Management Protocol* (SNMP) and can be tied into higher-level management platforms such as Hewlett-Packard's OpenView.

Among the capabilities of these wireless managers is the support of remote reboot, group configuration, or group software uploads for all the wireless infrastructure devices on the network. In addition, the LAN administrator can see how many client devices are connected to each AP, monitor those connections to measure link quality, and monitor all the APs for performance. Many of these management systems are vendor specific and are useful for managing an entire WLAN infrastructure comprised of their products.

The CiscoWorks *Wireless LAN Solution Engine* (WLSE), for example, provides centralized template-based configuration with user-defined groups to manage a large number of APs and bridges from Cisco. It monitors Cisco's *Lightweight Extensible Authentication Protocol* (LEAP) authentication server and enhances security management by detecting misconfigurations on APs and bridges that would allow intrusions. The functionality within CiscoWorks WLSE includes proactive monitoring, troubleshooting, the notification of performance degradation, and capabilities to improve capacity planning.

A different way of controlling multiple APs is promoted by Proxim, which offers the Harmony AP Controller. Traditional APs provide filtering and management, as well as radio functions, which inflates the cost of each device. The AP Controller separates management and filtering capabilities from the radio itself, providing real-time management and filtering for multiple APs. This results in physically smaller APs that can be placed wherever wireless coverage is needed. The controller supports up to 30 APs and provides a comprehensive systemwide view of all the devices through a web-browser-based tool. With the controller providing management, communication, and security services for all APs, an enterprise can achieve significantly lower costs on wireless infrastructure.

The APs are connected to the controller via standard Ethernet cabling. A low overhead IP-based communication protocol is used between the controller and the APs. Through this protocol, all APs receive their configuration settings from the controller. An optional power system can be used to supply power to the APs over the Ethernet cable at a distance of 330 feet (100 meters). This arrangement is referred to as *Power over Ethernet* (PoE) (see Chapter 3, "Setting Up the Network").

There is a downside to connecting all APs to a controller—if the controller fails, all clients will disconnect from the network. To guard against this possibility, Proxim recommends having at least one backup AP Controller in the network so if a controller fails, the APs associated with it will automatically be joined to another controller with the same system name. Because the controller is independent of radio type, it can support multiple APs of different standards (802.11a and 802.11b) at the same time.

Wireless Print Server

A wireless print server provides an easy solution for attaching a printer to a wireless network. Instead of risking printing bottlenecks by attaching a printer to one of the computers on the network, the printer can be placed anywhere in the range of other wireless devices and in a more convenient location accessible to all users. A wireless print server overcomes another downside to printer sharing in that the PC to which a printer is attached must be turned on in order for other computers to send print requests. There is also a performance hit on the host computer while other PCs are using the attached printer.

The wireless print server looks much like an AP, except that it hooks up to a printer with a parallel cable. The antenna provides the wireless link to desktop computers, notebooks, and handheld devices equipped with WiFi cards. The printer server has an RJ45 jack for an Ethernet connection, which is used to configure the device through a web browser. Print jobs are controlled from each user's computer or by the network administrator with a remote management utility.

By enabling flexible office layouts and easy moves, wireless print servers are an ideal solution for dynamic environments. By going wireless, companies are able to avoid costly moves and network modifications to relocate printers from one area to another. Employees do not need to consider such configuration problems when putting in new workgroups. This gives companies more flexibility to do anything they want without having to worry about physical constraints.

Wireless Bridges

Bridges are used to extend or interconnect LAN segments, whether the segments consist of wired or wireless links. At one level, they are used to create an extended network that greatly expands the number of devices and services available to each user. At another level, bridges can be used for segmenting LANs into smaller subnets to improve performance, control access, and facilitate fault isolation and testing without impacting the overall user population.

Basically two types of wireless bridges are available: workgroup and building to building. A workgroup bridge is used to meet the connectivity needs of remote workgroups, satellite offices, and mobile users. It connects to a hub along with multiple Ethernet-enabled notebooks or other portable computers. The bridge makes the wireless connection to an AP or another wireless bridge, which ultimately gets the traffic from the workgroup to the corporate LAN (see Figure 2-11). The bridge can also be used to implement centralized security and management features.

Some wireless bridges include features to deal with the multipath propagation often found in office environments. This occurs when the RF signal takes different paths to its destination, such as an AP. Along the way, the signal can encounter objects that cause it to bounce in different directions, causing a portion of the signal to be delayed in reaching the destination, which the user experiences as lower throughput.

Diversity antennas are an effective way of overcoming the effects of multipath propagation. This refers to the use of two antennas on the bridge to increase the odds of receiving a better signal on either of the antennas. The

Figure 2-11 A wireless workgroup bridge connects a cluster of remote users to the corporate LAN.

antennas are physically separate from the radio and they are adjustable, ensuring that one of the antennas will encounter a better signal than the other. The receiver uses signal-filtering and decision-making software to choose the best signal for demodulation. If range is a problem, higher-gain antennas may be used to extend the reach of the signal.

For the authentication of devices attached to the workgroup, a username and password may be stored in the workgroup bridge. When authenticated, the workgroup bridge receives a single-session, single-user encryption key from the RADIUS server via the associated AP. The wireless link between the bridge and AP is protected with WEP encryption.

Other bridge features include advanced diagnostic tools to simplify troubleshooting, remote system configuration, and management via the use of a browser, Telnet, *File Transfer Protocol* (FTP), or SNMP. The use of browsers is a relatively recent trend in network management. It lets network administrators link into a bridge or AP to perform various chores from any computer they happen to be near, instead of being tied to their desk all day long.

Telnet can also be used during the initial configuration of the device to set it up for operation as a bridge or AP and then later for other purposes such as running utilities, starting and stopping services, and reviewing log files. FTP can be used to update the firmware of a bridge or AP from an FTP server. SNMP is used to set various traps and send a message of their occurrence to a management station. Traps can be set to check for performance anomalies, security problems, equipment failures, compatibility issues, and other potential problems so that corrective action can be taken.

Workgroup bridges can be linked together to provide coverage to all of a company's offices within a building. Many small to medium-size businesses are deploying WiFi to complement an existing wired network, whereas others are turning to WiFi as an alternative to a wired network, thereby eliminating the expense and delay of installing CAT 5 cabling, particularly in older buildings that have been renovated for business tenants. If the company has to move to another building, the wireless bridges can be moved as well, preserving the capital investment.

Building-to-Building Bridges

Wireless bridging solutions span the distances between multiple buildings, across roads and highways, and even to the remotest of locations such as oil rigs or firefighting outposts. In cases where expensive leased lines are used for interbuilding connections, WiFi bridges offer the bandwidth of multiple

T1 lines, but without the cost of these lines, which can cost hundreds or thousands of dollars per month, depending on the distance. Whereas a T1 line can take up to 45 days to deliver from a local phone company, a wireless link can be deployed within a day or two. Although bridging solutions are an established tool of any 802.11b wireless network, organizations have found the previous wireless solutions limited due to vendors' use of proprietary protocols.

Enhanced bridging solutions are now available that are compatible with any WiFi-certified AP, easily fitting within a multivendor wireless infrastructure. This enables organizations to link multiple buildings in a campus over a greater distance and in a simpler way, while taking advantage of security and network management features.

Such bridges offer a cost-effective alternative to laying cable and can quickly provide connections between temporary locations such as trailers in construction sites or temporary classrooms in educational settings. They have an effective range of up to 15 miles at 11 Mbps. Organizations can set up the wireless bridge in point-to-point and point-to-multipoint configurations covering distances of 300 to 20,000 meters depending on the power of the antenna used.

To keep wireless information secure, these long-distance bridges support WEP encryption. They are also transparent to VPN protocols, using a special encapsulation protocol to avoid conflicts with other tunneling protocols.

When paired with Ethernet switches, administrators can set up traffic prioritization parameters to optimize data flow between buildings. This *quality of service* (QoS) feature enables the wireless bridge to identify and prioritize network traffic based on the application to ensure that business-critical traffic reaches its wireless destination within a specified time. Network administrators can configure wireless devices and also distribute software upgrades simultaneously to multiple wireless devices. Web-based management lets administrators choose between managing all bridges locally from any point on the subnet or remotely via the Internet.

Wireless Routers

Whereas APs move IP traffic between wireless clients and the wired LAN, wireless routers are used to create point-to-point and point-to-multipoint wireless links to connect buildings without the need to use cable or lease lines from a carrier. Routers are an effective way to provide blanket coverage to an entire campus, such as universities, hospitals, or office parks. They can also be used to connect school districts and municipalities to network buildings that are miles apart. They can even reach mobile notebooks, remote point-of-sale terminals, and vehicles such as police cars and fork-lifts. The range of the wireless links depends on the power of the antenna, but a range of 16 miles (26 km) is supported out of the box by some vendors, which is enough to provide metro area coverage.

Although some vendors claim that their products can support an unlimited number of users, this assumes that not all users will be active on the network at the same time. Performance decreases as the number of users increases, similar to the way throughput decreases on a wired network when a large number of users share the available bandwidth. Another issue that affects the number of users is the operating range of the wireless router. At a maximum distance of 16 miles (26 km), for example, the available bandwidth of a 2.4 GHz wireless connection might be reduced to only 1 Mbps, leaving much less bandwidth for each user.

Once the routers are deployed, the amount of bandwidth available on the individual wireless links can be determined. To ensure that users get the bandwidth they need, the network administrator can use a bandwidth management tool to limit each client to a specific increment of bandwidth for his or her own use such as 64, 128, 256, 384, or 512 Kbps. When bandwidth is not in use by some clients, other clients can use it. As more users come online, the clients fall back to their originally assigned increment of bandwidth.

Wireless routers for enterprise deployments include a number of other management features such as the following:

- **SNMP remote management** This enables a network administrator to monitor all the routers on the wireless network by setting traps for various operational and security anomalies and issuing alerts to a management station when they occur.
- ***Dynamic Host Configuration Protocol* (DHCP)** This feature enables computers to obtain their IP address, subnet mask, default router, domain name, and *Domain Name Server* (DNS) directly from the router. At the same time, the wireless routers can obtain their own IP address information from a DHCP server connected to the wired network.
- ***Network Address Translation* (NAT)** This feature enables multiple computers to share a single IP address to connect to the Internet. This enables the network administrator to assign corporate devices private IP addresses for internal networking without having to assign a limited number of registered IP addresses to every device to

make them recognizable on the public Internet. That way, the limited supply of IP addresses can be more wisely used for publicly available servers.

In addition to IP address conservation, NAT serves as a simple firewall for incoming connections since only traffic initiated by interior computers is permitted through the NAT mechanism. When the router receives a packet from a client device on the corporate network bound for a destination on the public Internet, the router rewrites the address so it appears to originate from one of its public IP addresses and the packet is sent out with this address.

When a reply packet comes back, it will be addressed to the public address previously assigned by the router. The router maintains a database of outstanding requests and will look up the address of the station that made the request. It then rewrites the public IP address of the return packet to the private IP address, ensuring its delivery to the appropriate client behind the router.

Both static and dynamic NATs are supported. Static address translations explicitly map an external address to an internal address. For incoming packets that have not been specifically requested, such as e-mail, static mapping is used. With dynamic translations, a pool is allocated and each new IP address to be translated is dynamically mapped to another IP address from the pool in a round-robin fashion.

NAT is not a foolproof security solution. Via address spoofing, the router can be tricked into giving access to an Internet intruder. Other measures, such as a firewall, must be taken to ensure more effective security. This can be a standalone device or an option available on the wireless router's operating system. The wireless link itself, however, is usually protected by WEP.

The capabilities, features, and configuration settings of the wireless router can be upgraded or changed by the network administrator from a remote management station using tools such as Telnet, FTP, and SNMP. Vendors typically offer Windows-based tools with their products to make configuration and management chores easier. Some offer policy-based management tools that enable administrators to easily configure a large number of wireless devices on the network. Under the policy-based approach, similar devices are grouped together into classes. When one device is configured, that configuration is carried over automatically to similar devices on the network. The policy approach can also be used to assign increments of bandwidth to different classes of users as well as stipulate which network applications have priority, when they have priority, and which users can access them.

Wireless Gateways

A wireless gateway is an application-specific device. A residential gateway, for example, is a type of AP that is used by telecommuters to provide WiFi-equipped computers at home with access to the Internet. This class of APs is tailored for the home market and usually does not have management features. In most respects, such devices operate in much the same way as APs.

A gateway for the enterprise, however, is rich in management features. An example of such a device is a security gateway, which may come in different versions for high-end enterprises, departments of up to 100 users, and smaller businesses or workgroups of up to 15 users. Such gateways are designed to offer a complete wireless networking solution consisting of the following:

- **Security** This is used for authentication and granular access control, enabling administrators to easily configure who can access their network and what resources or applications they can use.
- **Wireless link encryption** This entails setting up VPN tunnels to ensure that wireless communications are protected. The tunnel terminates directly on the wireless gateway.
- ***Class of service*** **(CoS)** This is used to assign each category of user with a set amount of bandwidth to control and maintain CoS.
- **Manageability** This is used to provide an intuitive web-based interface that facilitates rapid installation, configuration, and remote management.

Users can be authenticated against a database stored locally on the wireless gateway or against RADIUS, *Lightweight Directory Access Protocol* (LDAP), Windows NT domain servers, or a Windows 2000 Active Directory. A single wireless gateway is capable of supporting a mix of locally and centrally authenticated users. Users who log onto a Windows domain (NT or 2000) can be logged onto the wireless gateway transparently. Other users are automatically directed to the sign-on page via their web browser, which sets up a *Secure Sockets Layer* (SSL) connection. After users are authenticated, they are assigned a role. Each role is configured with predefined access rights that are enforced on that user session.

The gateway attaches to a network of wireless APs on one side and a company's wired LAN on the other. The device can offer a viable security alternative to VPNs. In general, an enterprise VPN views the wireless network as a potentially hostile network, as it views the Internet. With an *IP*

Security (IPSec) solution, the wireless clients must authenticate before they connect with the wireless network. VPNs can have a substantial administrative overhead. For example, VPN software has to be installed on all wireless clients, whereas a server program runs on one or more dedicated servers.

By contrast, the security gateway acts as a funnel, collecting all traffic from the AP and applying a broad range of security controls, including authentication (with existing RADIUS, LDAP, Active Directory, and Windows domain databases), access control, and encrypted sessions—whatever the organization happens to be using. The security gateway also has CoS features to sort traffic into different categories of importance. This enables employees, managers, visitors, and business partners to be assigned different degrees of access to the corporate network.

Conclusion

WiFi networks are progressing in sophistication very quickly in terms of operating range, security, management tools, and traffic-handling capabilities that will make it possible to eventually use wireless links for *voice over IP* (VoIP) and streaming applications. In terms of speed, efforts are under way to merge the IEEE 802.11a standard for 5 GHz wireless networks favored in the United States with the HiperLAN2 standard favored in Europe into a single standard known as the *5 GHz Unified Protocol* (5-UP). By tying multiple channels together, this standard could offer up to 108 Mbps of bandwidth, which is double the 54 Mbps available under the 802.11a standard. With such high data rates, in addition to guaranteed QoS and tight security, which are also under development, 5-UP could pose a real challenge to both 3G and wired networks.

One wireless component manufacturer, Atheros Communications, sees 5-UP as providing a single network that is capable of supporting both constant bit rate and bursty network traffic while also providing interoperability between low and high data rate devices. Next-generation wireless technology based on 5-UP will also enable new applications and services that either are not feasible or are difficult to implement using wired connections. Examples include integrated wireless monitoring and security systems, personal e-mail pads, portable electronic newspapers and books, portable music and video on demand, cordless Internet telephony, and utility meter reading.

CHAPTER 3

Setting Up the Network

Before computers can take advantage of WiFi connectivity to share files, printers, and other resources, including a high-speed Internet access connection, they must be configured for networking. In the Windows environment, computers running Windows 95/98/ME must be manually configured. Computers running Windows NT/2000 are usually configured by the IT department before being issued to employees. Windows XP, however, comes with integral support for WiFi, which means that many of the network configuration chores are done automatically without needing to install drivers and enter detailed information screen after screen. Once the computers are properly configured, regardless of what version of Windows they might be running, they will be accessible to each other over the network. With the addition of software, even Apple computers can share the same wired *local area network* (LAN) and wireless *access point* (AP) connections.

Although WiFi promises wireless connectivity within a building, campus, or region, cabling is still occasionally needed. If the wired LAN is being extended via an AP or bridge, for example, then these devices must be connected to the LAN with standard *Category 5* (CAT 5) cable. Often, one or more APs will be cabled to a hub or Ethernet switch so mobile users can connect with other users on the wired LAN. If the APs are connected to a switch that supports *Power over Ethernet* (PoE), AC power can be conveyed to the APs over the unused pairs within the CAT 5 cable, eliminating the need to position them near a power outlet. If the switch does not support PoE, then a separate device called a *power injector* can be used, which is colocated with the switch in a telephone closet.

Types of Networks

Two types of networks are available: client server and peer to peer. Both are used in the corporate environment, and wireless connections using WiFi-compliant equipment are easily accommodated to extend connectivity to notebook computers and other portable devices. Telecommuters, however, will usually have a peer-to-peer network. Not only is a client-server network unnecessary, but the peer network is cheaper and easier to set up, and requires little technical expertise to troubleshoot and administer.

Client-Server Network

In a client-server network, an application program is broken out into two parts, appropriately called the *client* and *server*, which exchange information over the network (see Figure 3-1).

The client portion of the program, or *front end*, is run by individual users at their computers and performs tasks such as querying a database, producing a printed report, or entering a new record. These functions are carried out through a database specification and access language known as *Structured Query Language* (SQL), which operates in conjunction with existing applications. The front-end part of the program executes on the user's workstation, drawing upon its *random access memory* (RAM) and *central processing unit* (CPU).

Sometimes the servers are used for routine office applications, which the clients can use on a first-come, first-served basis. Before users are granted access to metered applications, the software inventory is checked to determine whether copies are available. If no copies are available, a status message is issued, indicating that all copies are in use. The user waits in a queue until a copy becomes available.

This software metering enables the network administrator to control the concurrent usage of each application. The network administrator can also choose to be notified when users are denied access to particular applications when all available copies are in use. This may identify the need to pay an additional license charge to the vendor so more users can access the application.

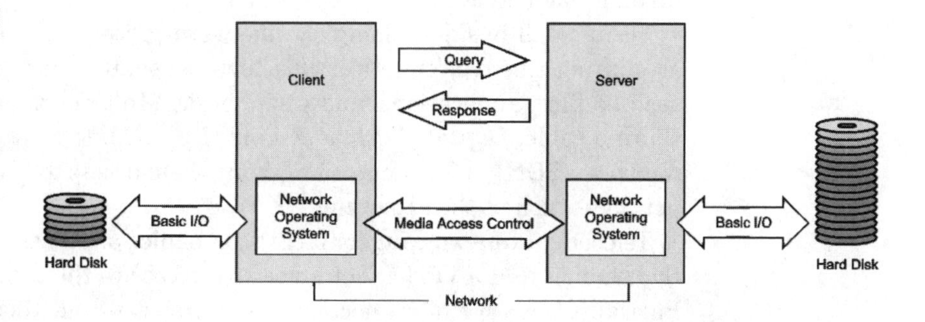

Figure 3-1 A simplified model of the client-server architecture

The server portion of the program, or *back end*, resides on a computer that is configured to support multiple clients, offering them shared access to numerous application programs as well as to printers, file storage, database management, communications, and other resources. The server not only handles simultaneous requests from multiple clients, but it also performs administrative tasks such as transaction management, security, logging, database creation and updating, and asset management.

The network consists of the transmission facility—usually the wired LAN. Among the commonly used media for LANs are coaxial cable (thick and thin), twisted-pair wiring (shielded and unshielded), and optical fiber (single mode and multimode). Increasingly, wireless links are being used to link clients and servers. When linking client-server computing environments over the wide area, other facilities and services come into play. T-carrier links provide bandwidth in a range of increments from 64 Kbps to 45 Mbps, whereas optical carrier links provide bandwidth in the multigigabit-per-second range. Carrier-provided services may also be used, such as IP-based *virtual private networks* (VPNs), frame relay, *Asynchronous Transfer Mode* (ATM), and metro-area Ethernet.

Peer-to-Peer Network

In a peer-to-peer network, computers are linked together through a hub or switch for resource sharing. They can assume the role of client or server at any given time. If a computer accesses the files, applications, or peripherals of another computer, it assumes the role of client and the other computer assumes the role of server. The next moment these roles can be reversed. Ultimately, the two computers have an equal relationship with each other, making them *peers*.

For a small business doing routine word processing, spreadsheets, and accounting, this type of network is a low-cost solution for sharing resources such as files, applications, and peripherals. Multiple computers can even share a cable, *Digital Subscriber Line* (DSL), *Integrated Services Digital Network* (ISDN), or T1 connection, giving them access to a high-speed Internet connection at the same time.

Telecommuters working for larger companies often use peer networks in their homes and a VPN to obtain secure access to the enterprise network. Basically, this type of connection lets users carve out their own IP-based *wide area network* (WAN) within the carrier's high-speed Internet backbone. Security functions, including encryption, are performed on the IP packets for routing across a tunnel through the Internet. By drawing on the larger Internet infrastructure, VPNs offer substantial cost savings over traditional private lines and data services such as frame relay.

WiFi can fit into the VPN scenario, but particular attention must be given to security because it is possible for hackers to find the wireless link and masquerade as a legitimate user to sneak into the enterprise network through the VPN tunnel. Because the average telecommuter does not have the expertise to deal effectively with security issues, many companies have adopted policies that forbid users from setting up WiFi links as a condition for being allowed to work at home or telecommute.

Larger companies with IT staff allow WiFi links in the home, but only after remotely configuring the user's computer for the VPN connection and walking the user through the encryption setup process that protects transmissions over the wireless link between the computer and AP. Most often, however, companies make the mistake of not specifically addressing the security issue of WiFi and count on the VPN to provide all the security they need.

Setting Up Computers for Networking

Whether computers are used in a peer-to-peer or client-server network, they must be configured for networking. In the corporate environment, computers are configured for networking by the IT staff before they are issued to employees. Generally, employees cannot change the configuration parameters or install new software programs without having administrator privileges. To qualify for administrator privileges, the employee must usually be in a technical position or be a power user who can lend his or her technical expertise to coworkers, thereby relieving the IT department of this handholding burden.

Windows 95/98/ME/XP and Windows NT/2000 can be used in either peer-to-peer or client-server environments. In addition, these versions of Windows support the same networking protocols—including the *Transmission Control Protocol/Internet Protocol* (TCP/IP) for accessing intranets, VPNs, and the public Internet—and provide options such as dial-up networking and fax routing.

One difference between the versions of Windows is that network connections are easier to set up in NT/2000 than in 95/98/ME/XP. With Windows

XP, network connection setup is part of the initial program installation. With Windows NT/2000 in a client-server environment, electronic software distribution and policy management capabilities enable IT managers to configure a large number of computers by invoking automated processes.

Windows supports Ethernet, token ring, ATM, *Fiber Distributed Data Interface* (FDDI), and other data frame types. Ethernet is by far the least expensive network to implement. The *network interface cards* (NICs) for each desktop computer cost as little as $20 each for 10/100 Mbps LANs, whereas a wireless NIC for a desktop computer costs less than $50. Most desktop and notebook computers purchased today include Ethernet capability as part of the standard configuration. If it is not included, Ethernet adapters are available for about $30, which plug into any available *universal serial bus* (USB) port. This simple solution enables an RJ45 Ethernet port to be added without opening the computer. WiFi adapters for desktop and notebook computers are inexpensive as well. For 2.4 GHz operation, wireless cards for desktop computers sell for less than $50, whereas those for notebooks sell for under $70. For 5 GHz operation, the cards for desktop and notebook computers cost between $100 and $200.

A five-port 10/100 Mbps hub costs as little as $40, whereas an eight-port model can be purchased for under $100. Although these are great for the peer-to-peer networks used by telecommuters, small businesses, and the branch offices of larger firms, an enterprise usually requires a high-end device that has more ports, is easily expandable, and offers management capabilities. Enterprise hubs cost $30 to $40 a port, but the price is often justified by the increased features and functionality. For example, the hub's management system can be used for creating *virtual LANs* (VLANs), which can be thought of as closed user groups within the greater LAN. Adding to the price is the hub's management module, which usually offers full *Simple Network Management Protocol* (SNMP) read/write configuration and traffic analysis as well as browser-based management from anywhere on the network. At an extra cost, support for *Remote Monitoring* (RMON) at the hub provides the information needed to easily track and troubleshoot network problems.

CAT 5 cabling usually costs less than 50 cents per foot in various lengths up to 100 feet with the RJ45 connectors already attached at each end. This cable is ideal for small installations where on-site technical assistance is not available. For large installations, cable can be purchased in rolls of 500 or 1,000 feet for even greater cost savings, but this does not include the RJ45 connectors. The presence of on-site technical support often makes the purchase of cable in bulk worthwhile for large organizations.

Configuration Details

When setting up a network, each computer can be configured individually or in large numbers remotely by a system administrator. If the computer does not already have an Ethernet card, one must be installed. Even though the computer is intended to be used over a wireless connection, the availability of a standard Ethernet port on the computer will facilitate the initial configuration of the AP, which is discussed in Chapter 7, "Wireless Security." Briefly, a cable connection between a computer and AP enables the user to set the configuration of the AP using a graphical interface. After the initial configuration, the cable is removed and the AP is cabled to a hub. The computer can then communicate with the AP over the wireless link and gain access to the LAN.

Furthermore, the availability of a standard Ethernet port gives the user more flexibility on where the computer is used. If a desktop computer is moved out of the range of an AP, for example, the only recourse might be to plug it into an available RJ45 jack. Likewise, if a notebook computer is used at a different location where no wireless connectivity exists, the standard Ethernet port might have to be used, and if that is not available, a dial-up connection would be the last resort. Mobile professionals, in particular, will appreciate a notebook computer that is fully equipped to take advantage of any connectivity option.

Install the Card If the computer does not come with Ethernet capability, a card must be installed while the computer is off. When powering up the computer with the card installed, Windows will usually recognize the new hardware and automatically install the appropriate network card drivers (see Figure 3-2). If the drivers cannot be found, Windows will prompt the user to insert the manufacturer's disk so the drivers can be installed.

IP Settings By selecting Internet Protocol (TCP/IP) and clicking on Properties, the user can set the IP address, subnet mask, default gateway, and *Domain Name Server* (DNS) server addresses (see Figure 3-3). In the corporate environment, IP addresses will likely be assigned to computers at the time they log onto the network from a *Dynamic Host Configuration Protocol* (DHCP) server. If they are not assigned, a static IP address must be entered, which is permanently assigned to the computer. The subnet mask will usually be set to 255.255.255.0. The default gateway is the IP address of the router, cable/DSL modem, or network appliance that provides access to the Internet.

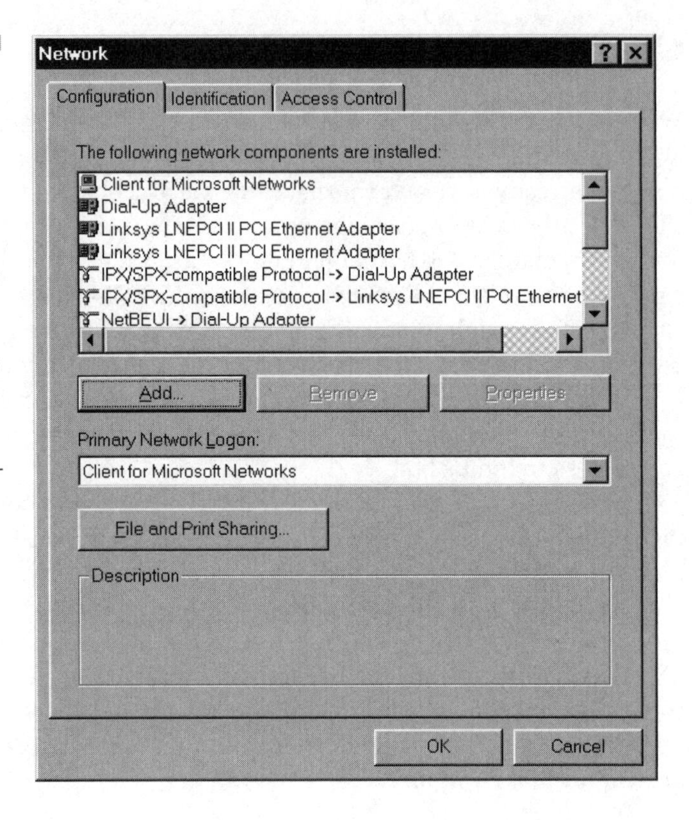

Figure 3-2 To verify that the correct drivers have been installed, the user opens the Network Control Panel to check the list of installed components. In this case, a Linksys LNEPCI II Ethernet Adapter has been installed.

Two DNS server addresses are required: one primary (preferred) and the other secondary (alternate). The DNS makes the translations between plain names, such as Amazon.com, and their assigned 32-bit IP addresses, such as 207.171.181.16. This translation capability makes it easy for users to navigate the Internet, so instead of memorizing 207.171.181.16, users only need to enter Amazon.com in the browser's command line. Two DNS addresses must be entered because if the primary DNS server fails, the secondary DNS server takes over.

Configure a Connection With Windows 95/98/ME, the user must go through a series of screens to properly configure the computer for networking. These configuration screens differ slightly depending on which version of Windows is being used, but they essentially prompt the user for the same information.

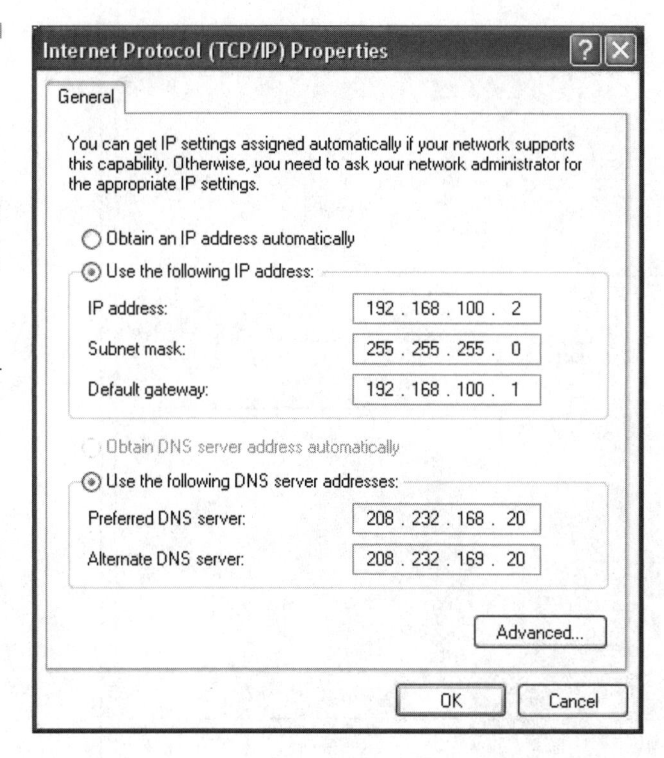

Figure 3-3 The Internet Protocol (TCP/IP) Properties screen provides the fields to enter IP address information, which enables the computer to access the Internet.

With Windows 98, for example, the user starts by specifying the client type. Because the network is based on Windows, the user must add Client for Microsoft Networks as the primary network logon (refer to Figure 3-2). Because the main advantage of networking computers is resource sharing, the user can choose to share both printers and files. The user does this by clicking on the File and Print Sharing button and choosing one or both of these capabilities. Through file and printer sharing, each computer becomes a potential server. However, the user may choose not to enable file sharing. This is common in enterprise networks, where shared directories and files reside on servers and are password protected according to the division, department, workgroup, project, or other scheme.

Identification From the Identification tab of the dialog box, the user selects a unique name for the computer and the workgroup to which it belongs, as well as a brief description of the computer (see Figure 3-4). When others use Network Neighborhood to browse the network, they will see all the active computers on the network.

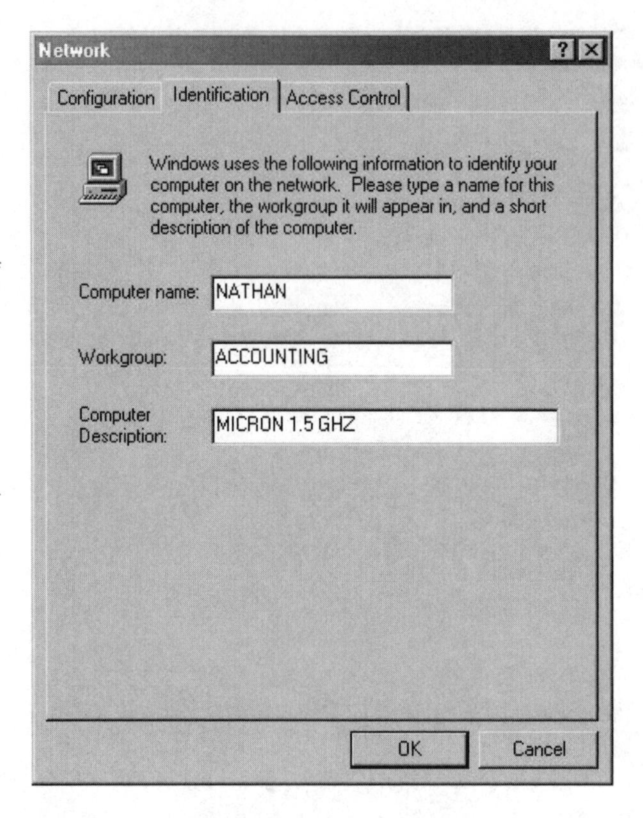

Figure 3-4 The user selects a unique name for the computer and the workgroup to which it belongs, and provides a brief description of the computer to identify it to other users when they access Network Neighborhood to browse the network.

Access Control From the Access Control tab of the dialog box, the user selects the security type. For a small workgroup, share-level access is adequate (see Figure 3-5). This enables printers, hard drives, directories, and other resources to be shared, and enables the user to establish password access for each of these resources. In addition, read-only access enables users to view, but not modify, a file or directory.

Printer Sharing To allow a printer to be shared, for example, the user right-clicks on the printer icon in the Control Panel and selects Sharing from the drop-down list (see Figure 3-6). Next, the user clicks on the Shared As radio button and enters a unique name for the printer (see Figure 3-7). If desired, this resource can be given a password as well. When another computer tries to access the printer, the user will be prompted to enter the password. If a password is not necessary, the password field is left blank, enabling anyone to use the printer.

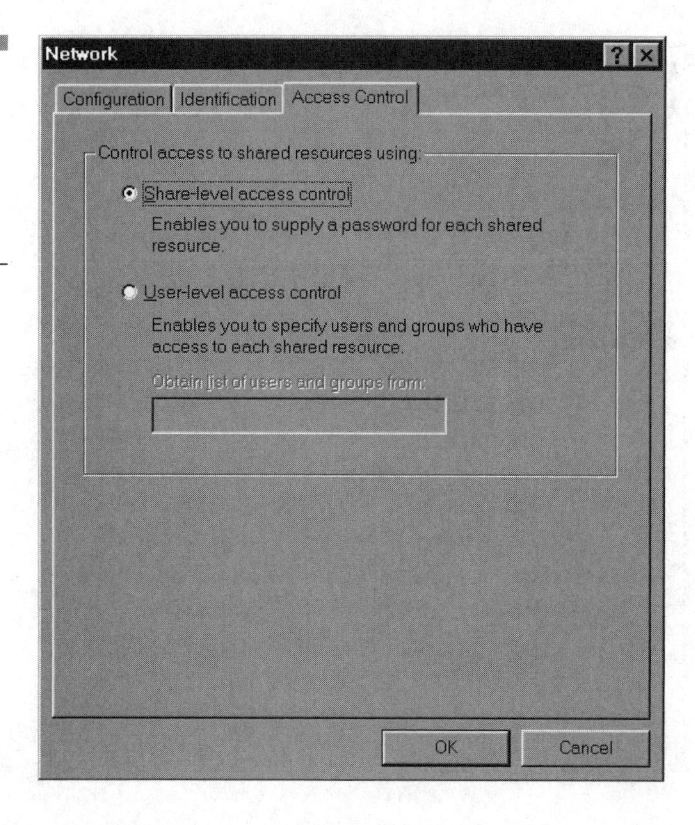

Figure 3-5 Choosing share-level access enables the user to password protect each shared resource.

In the enterprise environment, printers usually have their own IP address and are cabled to a hub or switch so it is not necessary to allow printer sharing unless a particular computer has a printer attached to it. Sometimes a workgroup within the enterprise may be set up this way. A graphics workgroup, for example, may have expensive high-resolution color scanners and printers for designing sales brochures and other materials. They do not want everyone in the company using these resources so they set up a peer network among themselves with the printers and scanners attached to particular workstations. Because these resources are directly attached to the workstations, they do not need IP addresses. They can still be shared by members of the workgroup.

Another security option in the Access Control tab is user-level access, which is used to limit resource access by username. This function eliminates the need to remember passwords for each shared resource. Each user simply logs onto the network with a unique name and password; the net-

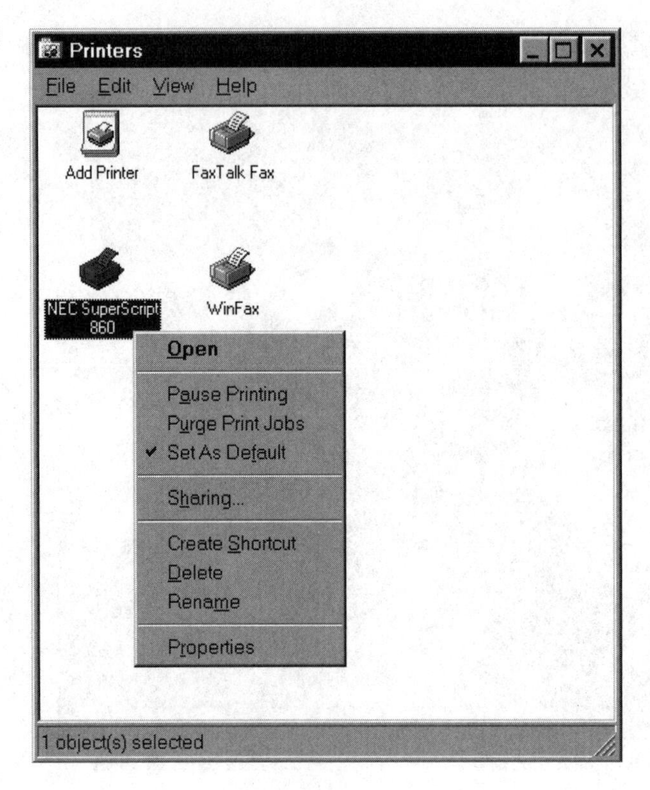

Figure 3-6 Any computer-attached printer can be configured for sharing.

work administrator governs who can do what on the network. However, this requires the computers to be part of a larger network with a central server —perhaps running Windows NT/2000 Server—which maintains the *access control list* (ACL) for the whole network. Because all versions of Windows support the same protocols, Windows 95/98/ME/XP computers can participate in a Windows NT/2000 server domain.

Peer services can be combined with standard client-server networking. For example, if a Windows 95/98/ME/XP computer is a member of a Windows NT/2000 network and has a color printer to share, the resource owner can share that printer with other computers on the network. The server's ACL determines who is eligible to share resources.

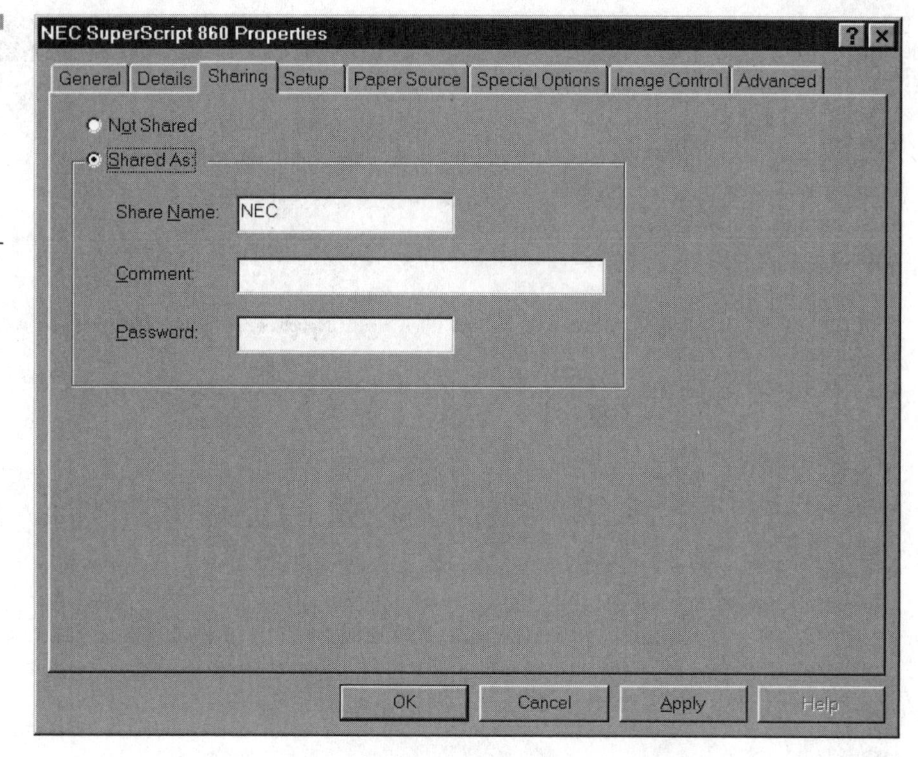

Figure 3-7 Like any other resource, access to any computer-attached printer can be controlled with a password.

Network Configuration with XP

Windows XP offers the Network Setup Wizard, making it easier to configure a connection by offering users the following choices (see Figure 3-8):

- This computer connects directly to the Internet. The other computers on My Network connect to the Internet through this computer (see Figure 3-9).
- This computer connects to the Internet through another computer on My Network or through a residential gateway (see Figure 3-10).
- This computer connects to the Internet directly or through a network hub. (Refer to Figure 3-7. This choice is revealed by clicking Other.) Other computers on My Network also connect to the Internet directly or through a hub (see Figure 3-11).

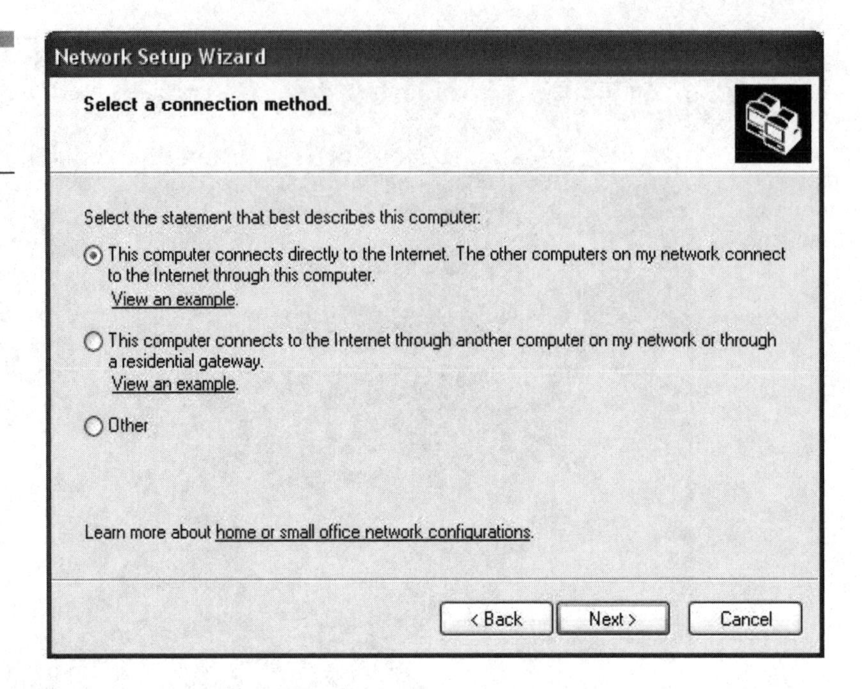

Figure 3-8 The Windows XP Network Setup Wizard

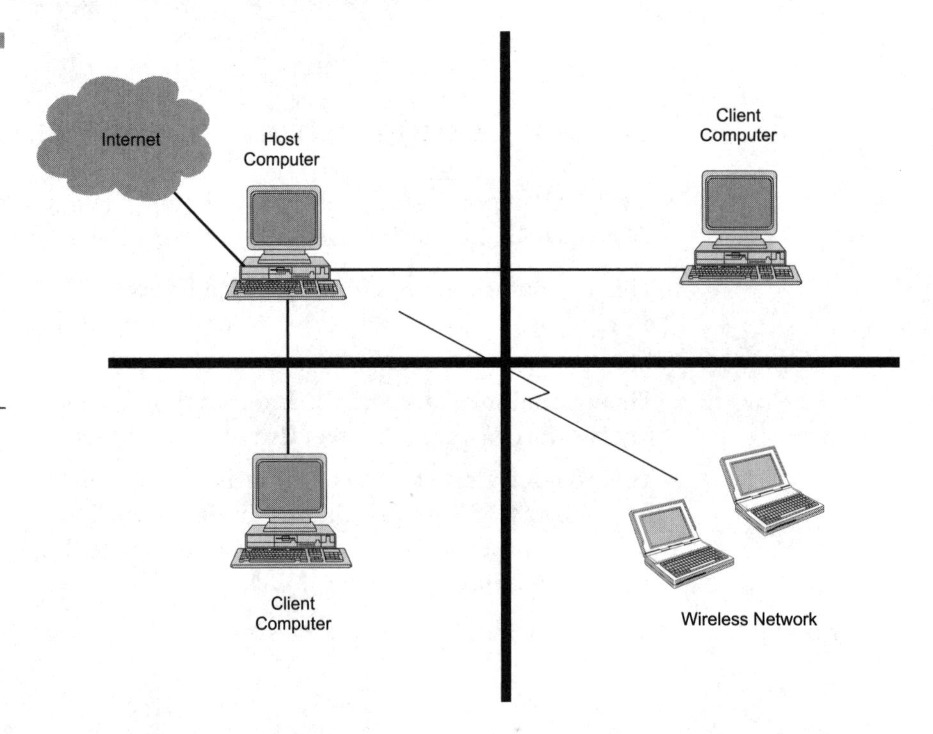

Figure 3-9 In this configuration, one computer is designated as the host through which all other computers are connected in order to gain access to the Internet.

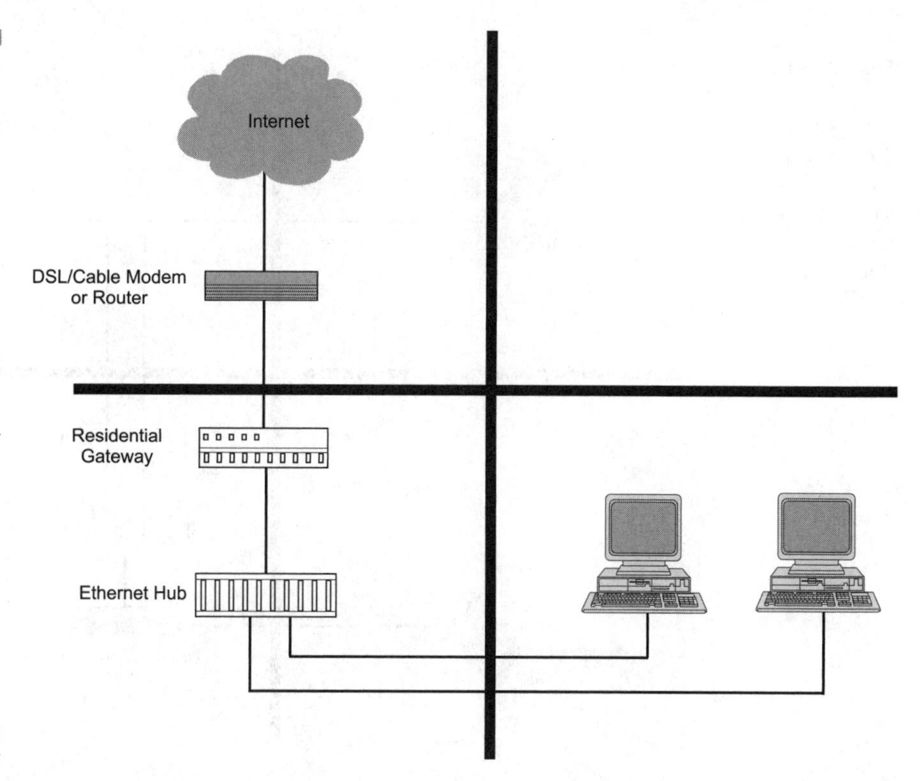

Figure 3-10 In this configuration, the computer is part of a home or small office network that connects to the Internet through another computer on the network or using a residential gateway.

After selecting the most appropriate connection method, the Network Setup Wizard suggests a connection that will be most likely used based on the user's input up to this point (see Figure 3-12). If LAN is selected, the Network Setup Wizard prompts the user for a computer description and computer name (see Figure 3-13), and then a workgroup name (see Figure 3-14), which identifies the network. The Network Setup Wizard summarizes the settings from all the previous configuration screens before applying them when the user clicks on the Next button (see Figure 3-15).

Firewall Windows XP comes with a built-in client firewall called the *Internet Connection Firewall* (ICF), which protects the organization's internal network from outsiders. It is used to set restrictions on what information is allowed to come into the network or go out of the network. If the computer has a VPN connection, however, the firewall should not be used because it will interfere with the operation of file sharing and other VPN functions. If the network is already protected by a firewall—at the router or some other network appliance, for example—the ICF is not needed.

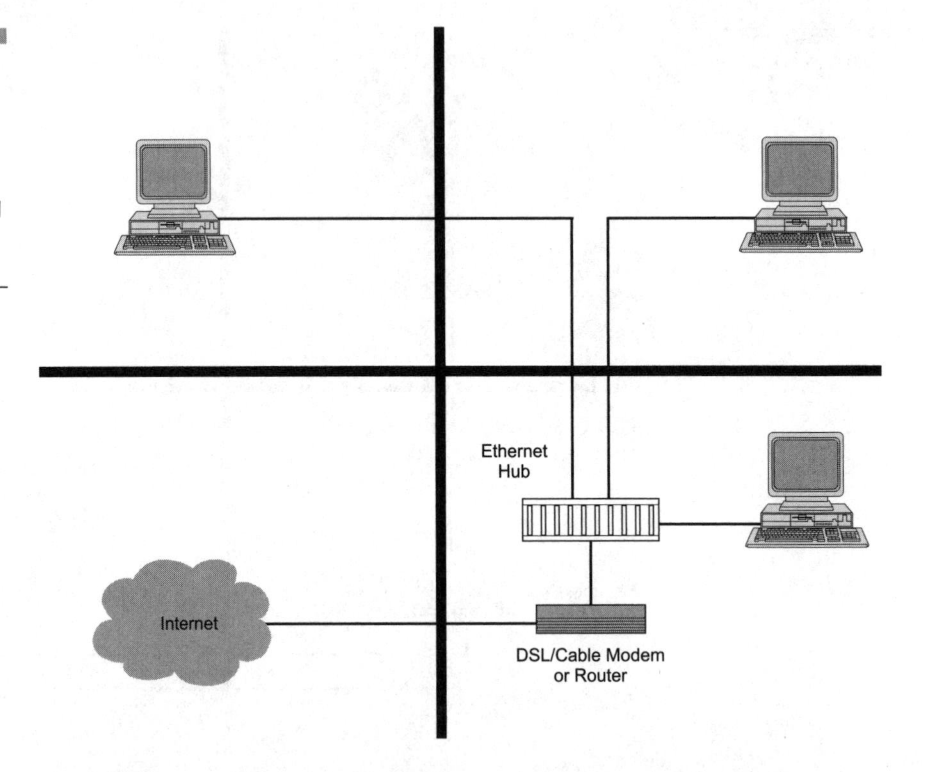

Figure 3-11 In this configuration, the computers are connected to each other and to the Internet through a hub.

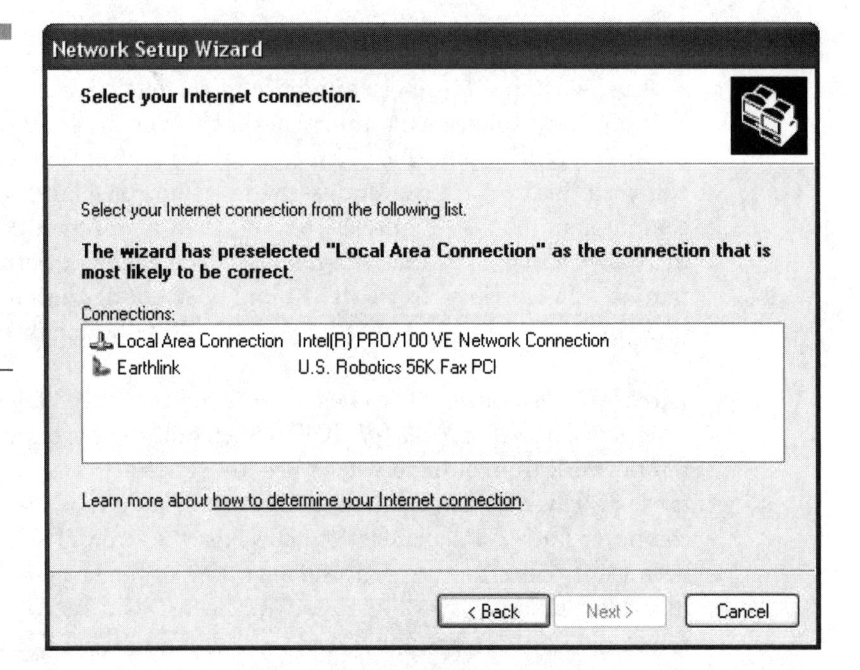

Figure 3-12 In this case, two possible connections are available to choose from—a LAN connection and a dial-up connection (previously configured).

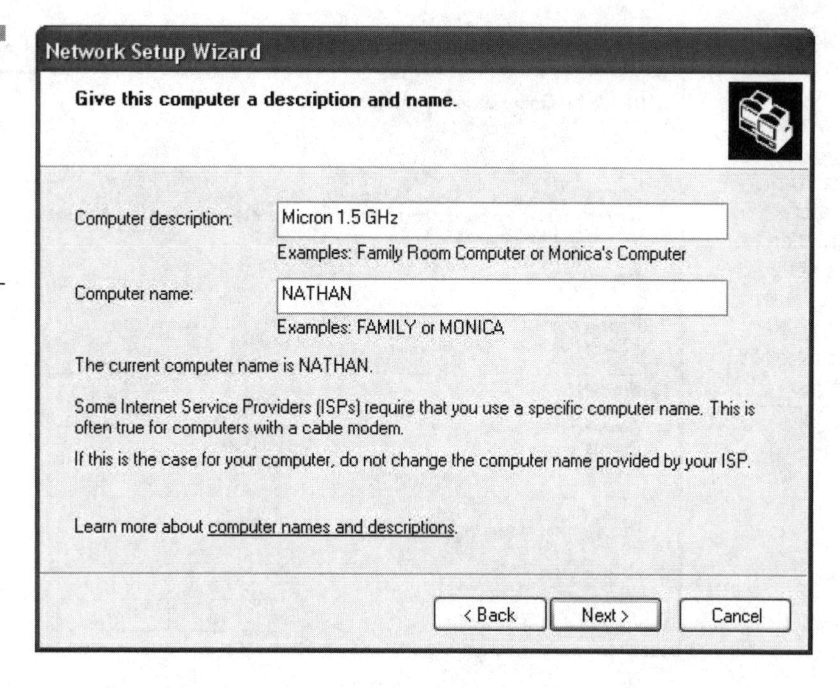

Figure 3-13 To configure the computer to work on a LAN, the user enters a computer description and a computer name.

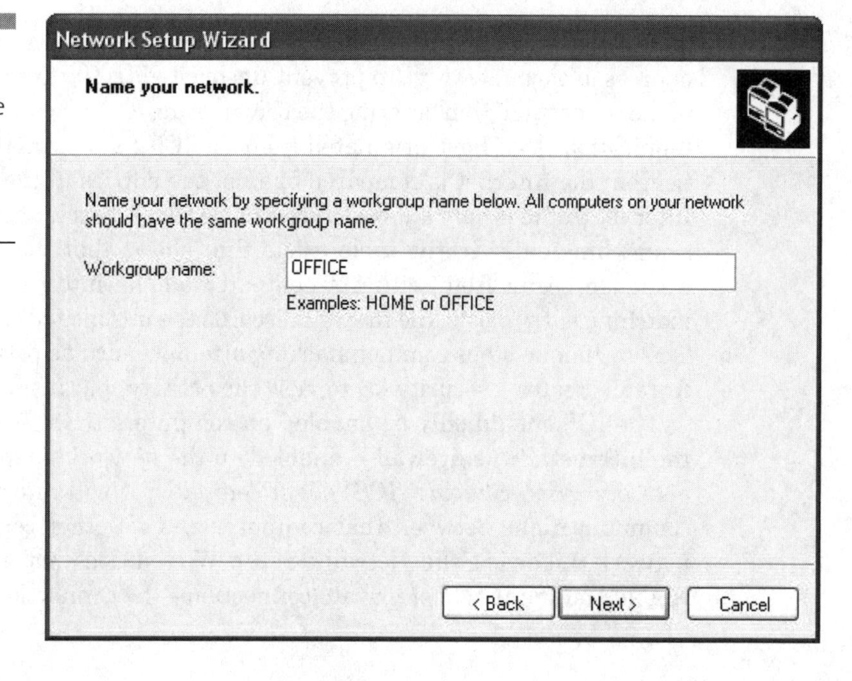

Figure 3-14 The same workgroup name must be used for all computers on the same network.

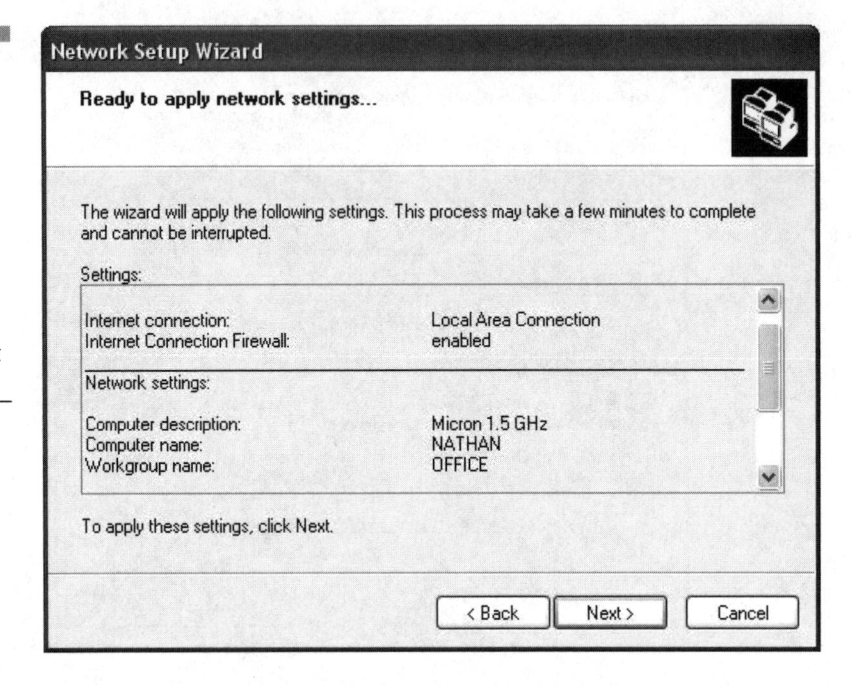

Figure 3-15 The Network Setup Wizard summarizes the settings from all the previous configuration screens before applying them when the user clicks on the Next button.

The ICF is a stateful firewall that inspects the source and destination address of each message. To prevent unsolicited traffic from the public side of the connection from entering the private side, ICF keeps a table of all communications that have originated from the ICF computer. All inbound traffic from the Internet is compared against the entries in the table. Inbound Internet traffic is only allowed to reach the computers in the network when a matching entry in the table exists that shows that the communication exchange began from within a computer on the internal network. If no matching entry exists, the firewall discards the unsolicited communications. This technique stops common hacking attempts such as port scanning. The firewall creates a security log to view the activity that it tracks.

The ICF should only be enabled on computers that connect directly to the Internet. If the firewall is enabled on the network adapter of an *Internet Connection Sharing* (ICS) client computer, it will interfere with some communications between that computer and all other computers on the network. Likewise, the Network Setup Wizard does not allow ICF to be enabled on the ICS host private connection—the connection that connects the ICS host computer with the ICS client computers—because enabling a firewall in this location would prevent network communications.

Cable Connection

Once the networking infrastructure is in place, the NIC of each computer is individually connected to a hub with CAT 5 cable. This cable has connectors on each end, which insert into the RJ45 jacks of the hub and NICs. For small networks, the hub will usually not have management capabilities. Once the computers are properly configured and connected to the hub, the network is operational.

Hub

With today's networks becoming increasingly more complex, the conventional bus and ring LAN topologies have exhibited shortcomings, especially with regard to cable installation and maintenance. Furthermore, a fault anywhere in the cabling usually brought down the entire network or a significant portion of it. This weakness was compounded by technicians' inability to readily identify the point of failure from a central administration point, which tended to prolong network downtime. This situation led to the development of the hub in the mid-1980s.

Hubs provide a central point at which all cables meet, including the cables from wireless APs. They are at the center of the star configuration, with the individual cables (that is, segments) radiating outward to connect the various network devices, which may include bridges/routers that connect to remote LANs via the WAN (see Figure 3-16). Wiring hubs physically convert the networks from a bus or ring topology to a star topology while logically maintaining their Ethernet (or token ring) characteristics. The advantage of this configuration is that the failure of one segment—which may be shared among several devices or dedicated to just one device—does not necessarily impact the performance of other segments.

Not only do hubs limit the impact of cabling faults to a particular segment, but they also provide a centralized administration point for the entire network. If the wiring hub also employs some CPUs and management software to automate fault isolation, network reconfiguration, and statistics-gathering operations, it is no longer just a hub—it is an *intelligent*

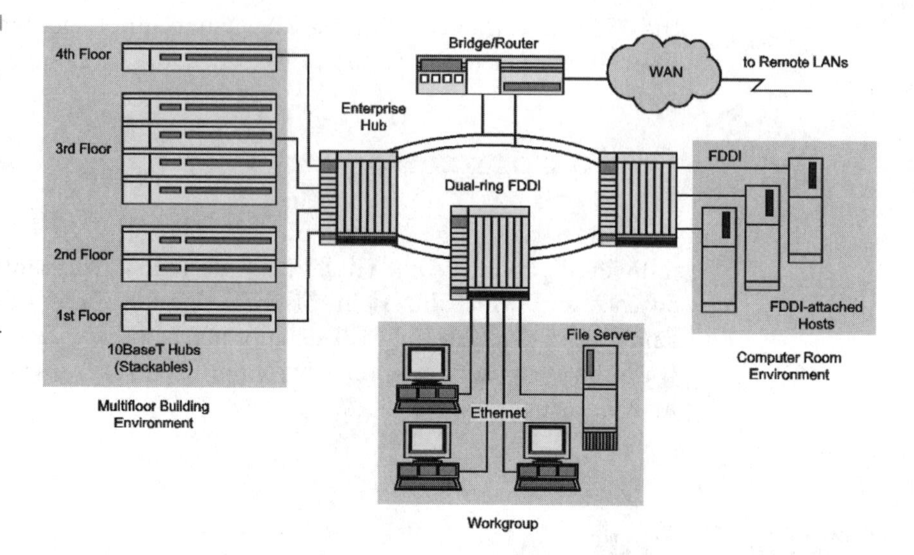

Figure 3-16 Some connectivity options available through various types of hubs serving multifloor building, workgroup, and computer room environments

hub that is capable of solving a wide range of connectivity problems efficiently and economically.

Types of Hubs

High-end hubs are modular in design, allowing the addition of ports, network interfaces, and special features as they are needed by the organization. These enterprise-level hubs can support networks that combine different LAN topologies and media types in a single chassis. Ethernet, token ring, and FDDI networks can coexist in a single hub. LAN segments using CAT 5 cable (that is, twisted-pair wiring), coaxial cable, and optical fiber also can be interconnected through the hub. In such cases, the hub is equipped with media-conversion modules to pass traffic from one type of medium to another.

Other hubs are available in fixed configurations for departments or workgroups that do not anticipate future growth. A variation of the fixed-configuration hub is the stackable hub. A unique feature of stackable hubs is that they can be interconnected through a modular backplane. This enables managers to economically expand workgroup and departmental networks as needed. Whereas the high-end modular systems are used to build large-scale enterprise networks, *stackables* are designed for small to medium-sized networks.

Workgroup hubs are also available, which typically have eight ports or less. This type of hub not only provides connectivity among connected workstations, but it also connects to a DSL or cable router to enable all the workstations to share the same Internet access connection. This type of hub is popular for small businesses, *small office/home office* (SOHO) environments, and telecommuters who have multiple PCs in their home.

A relatively new category of hub is the *superhub*. Superhubs are modular units that provide at least an uplink to a standalone ATM switch, if not some level of integral ATM switching, in addition to 10/100/1000 Mbps LAN support, and integrated LAN switching and routing. Fully populated superhubs support in excess of 500 ports of mixed-media, shared, and switched connectivity over a redundant gigabit-per-second backplane in a software-manageable, fault-tolerant, hot-swappable modular chassis that can cost well over $100,000.

Hub Components

Enterprise-level intelligent hubs contain four basic components—a chassis, a backplane, plug-in modules, and a network management system.

Chassis The chassis is the hub's most visible component. It contains an integral power supply and/or primary controller unit and varies in the number of available module slots. The modules insert into the chassis and are connected by a series of buses, each of which may constitute a separate network or integrate with one or more backbone networks. The chassis holds the individual modules. To fit into the chassis, each module is instantly connected to other modules via the hub's high-speed backplane.

Backplane The main artery of the hub is its backplane, a board that contains one or more buses that carry all communications between LAN segments. The hub's backplane is analogous to a PC bus through which various interface cards may be interconnected. The data path that carries traffic from card to card is often called a *channel*; unlike the PC, though, the hub's backplane typically consists of multiple physical or logical channels. The hub accommodates at least one LAN segment for each channel on the backplane.

Segmenting the backplane in this way enables multiple independent LANs or LAN segments to coexist within the same chassis. A separate backplane channel is usually present to carry management information. The segmented backplane typically has dedicated channels for Ethernet, token ring, and FDDI networks. Some hubs employ a multiplexing technique across the backplane to divide the available bandwidth into multiple logical channels. Other hubs support load sharing that enables network modules to select the backplane channel that will transport the traffic. Still other hubs are designed to enable backplanes to be added or upgraded to accommodate network expansion and new technologies.

The potential bandwidth capacity of newer backplane designs supporting ATM switching is quite impressive, reaching well into the gigabit-per-second range—more than enough to accommodate several Ethernet, token ring, and FDDI networks simultaneously.

Modules The functionality of hubs is provided by individual modules. The type of module used depends on the hub vendor. Typically, the vendor will provide multiuser Ethernet and token ring cards, LAN management, and LAN bridge and router cards. The use of bridge and router modules in hubs overcomes the distance limitations imposed by the LAN cabling and facilitates communication between LANs and WANs.

Plug-in modules are even available for terminal servers, communications servers, file and application servers, and *Systems Network Architecture* (SNA) gateways. Hub vendors also offer a variety of WAN interfaces, including those for IP, frame relay, ISDN, T-carrier, and ATM. As many as 60 different types of modules may be available from a single hub vendor, many of them provided under a third-party *original equipment manufacturer* (OEM), technology swap, and other vendor-partnering arrangements.

Modules plug into vacant chassis slots. Depending on the vendor, the modules can plug into any vacant slot or slots specifically devoted to their function. Hubs supporting any-slot insertion automatically detect the type of module that is inserted into the chassis and establish the connections to other compatible modules. In addition, many vendors offer a hot-swap capability that permits modules to be removed or inserted without powering down the hub.

Management System Hubs occupy a strategic position on the network, providing the central point of connection for workstations, servers, hosts, bridges, routers, and APs on the LAN and over the WAN. The hub's management system is used to view and control all devices connected to it, providing information that can greatly aid troubleshooting, fault isolation, and administration. The management tools typically fall into five categories: accounting management, configuration management, performance management, fault management, and security management.

Hub vendors typically provide proprietary management systems that offer value-added features that can make it easier to track down problem-causing workstations or servers. Most of these management systems support SNMP, enabling them to be controlled and managed through an existing enterprise management platform such as Hewlett-Packard's OpenView. Some hubs have RMON embedded in the hub, making possible more advanced network monitoring and analysis up to *Open Systems Interconnection* (OSI) Layer 7 (the application layer).

In summary, hubs are now the central point of control and management for the elements that make up departmental and enterprise networks. Hubs, which were developed to simplify the management of structured wiring as networks became bigger and more complex, enable the wiring infrastructure to expand in an orderly and cost-effective manner as the organization's computer systems grow and move and as interconnectivity requirements become more sophisticated.

Cabling

Twisted-pair wiring is the most common transmission medium; it is installed in office buildings and residences for telephone service, and because it is CAT 5 cable, it also supports high-speed data. Twisted-pair wiring consists of pairs of copper wires. To reduce cross talk or electromagnetic induction between pairs of wires, two insulated copper wires are twisted around each other. For some business locations, twisted pair is enclosed in a shield that functions as a ground. This is known as *shielded twisted pair* (STP). Ordinary wire is called *unshielded twisted pair* (UTP).

The same UTP wiring has become the most popular transmission medium for local area networking (see Figure 3-17). The pairs of wires in UTP cable are color coded so that they can be easily identified at each end. The most common color scheme is the one that corresponds to the *Electronic Industry Association/Telecommunications Industry Association's* (EIA/TIA's) Standard 568-B. Table 3-1 summarizes the proper color scheme.

The cable connectors and jacks that are most commonly used with CAT 5 UTP cables are RJ45. RJ simply means *registered jack* and the number 45 designates the pin numbering scheme. The connector is attached to the cable and the jack is the device that the connector plugs into, whether it is in the wall, in the computer's NIC, or in the hub.

Figure 3-17
CAT 5 UTP cables typically contain four pairs made up of a solid color and the same solid color striped onto a white background.

Table 3-1

Color scheme specified by EIA/TIA's Standard 568-B for CAT 5 UTP cable

Pin #	Signal	Wire Color	Used/Unused
1	Transmit +	White/orange	Used
2	Transmit −	Orange/white or solid orange	Used
3	Receive +	White/green	Used
4	N/A	Blue/white or solid blue	Not used
5	N/A	White/blue	Not used
6	Receive −	Green/white or solid green	Used
7	N/A	White/brown	Not used
8	N/A	Brown/white or solid brown	Not used

In response to the growing demand for data applications, cable has been categorized into various levels of transmission performance, as summarized in Table 3-2. The levels are hierarchical in that a higher category can be substituted for any lower category.

Table 3-2

Categories of UTP cable

Category	Maximum Bandwidth	Application	Standards
7	600 MHz	1000BaseT and faster	Standard under development *
6	250 MHz	1000BaseT	TIA/EIA 568-B (CAT 6)
5E	100 MHz	Same as CAT 5 plus 1000BaseT	ANSI/TIA/EIA 568-A-5 (CAT 5E)
5	100 MHz	10/100BaseT 100 Mbps *Twisted Pair Distributed Data* Interface (TPDDI) (ANSI× 319.5) 155 Mbps ATM	TIA/EIA 568-A (CAT 5) *National Electrical Manufacturers Associations* (NEMA) (extended frequency) ANSI/*Insulated Cable Engineers Assocation* (ICEA) S-91-661
4	20 MHz	10 Mbps Ethernet (IEEE 802.3) 16 Mbps token ring (IEEE 802.5)	TIA/EIA 568-A (CAT 4) NEMA (extended distance) ANSI/ICEA S-91-661
3	16 MHz	10 Mbps Ethernet (IEEE 802.3)	TIA/EIA 568-A (CAT 3) NEMA (standard loss) ANSI/ICEA S-91-661
2	4 MHz	IBM Type 3 1.544 Mbps T1 1 Base 5 (IEEE 802.3) 4 Mbps token ring (IEEE 802.5)	IBM Type 3 ANSI/ICEA S-91-661 ANSI/ICEA S-80-576
1	Less than 1 MHz	*Plain Old Telephone Service* (POTS) RS 232 and RS 422 ISDN Basic Rate	ANSI/ICEA S-80-576 ANSI/ICEA S-91-661

* In new installations, fiber to the desk may be less expensive than installing CAT 7 cable.

UTP cable has evolved over the years, and different varieties are available for different needs. Improvements over the years, such as variations in the twists or in the individual wire sheaths or overall cable jackets, have led to the development of EIA/TIA-568 standard-compliant categories of cable that have different specifications on signal bandwidth. Because UTP cable is lightweight, thin, and flexible, as well as versatile, reliable, and inexpensive, millions of nodes have been and continue to be wired with UTP cable, even for high data rate applications. For the best performance, UTP cable should be used as part of a well-engineered and structured cabling system. However, businesses that require reliable gigabit-per-second data transmission speeds should give serious consideration to moving to optical fiber rather than CAT 7 UTP.

Power over Ethernet (PoE)

Some business-class WiFi APs can be configured to derive their power from the CAT 5 cable connection, instead of requiring that they be positioned near a power outlet so they can be plugged in. This feature, PoE, simplifies and reduces installation and operation costs for large organizations. Electrical power comes from an alternative source, such as a powered switch, a powered patch panel, AP controller (that is, Proxim's Harmony AP Controller), or a small inline device called a *power injector*. The reason why power can be run through a segment of CAT 5 cabling is because only four of the eight wires inside the cable are actually used for sending and receiving data traffic. The other four wires are idle and can be used for other purposes. (Refer to Table 3-1.)

In the case of a power injector, the device is installed in the wiring closet near the Ethernet switch or hub and inserts a DC voltage onto the CAT 5 cable, which runs the AP (see Figure 3-18). Some APs and other network

Figure 3-18 SMC's Power Injector is a PoE device that provides power to an AP via a CAT 5 Ethernet cable, reducing installation and labor costs by eliminating the need to install the AP close to a power outlet.

devices accept the injected DC power directly from the CAT 5 cable through the RJ45 jack. These devices are considered to be PoE compatible or active Ethernet compatible.

Two types of injectors are available: passive and fault protected. Each type is typically available in a variety of voltage levels and number of ports. Passive injectors place a DC voltage onto a CAT 5 cable. These devices do not provide short-circuit or over-current protection. Fault-protected injectors provide continuous fault monitoring and protection against any detected short circuits and over-current conditions in the CAT 5 cable.

During normal operation a fault should never occur in the CAT 5 cable, but a fault might be introduced into the cable in a few ways. For example, the attached device may be totally incompatible with PoE and may have some nonstandard or defective connection that short circuits the conductors normally used for PoE. Currently, however, most non-PoE devices have no connection on the PoE pins. In addition, cabling that has its individual wires connected differently at both ends can cause a fault. If the insulation becomes damaged to cause contact between wires inside the cable, this can cause a fault as well. This can happen if the cable is cut, crimped, or crushed during installation or follow-on construction work.

If a fault occurs, the fault-protection circuit shuts off the DC voltage injected onto the cable. Fault-protection circuit operation varies according to model. Some models continuously monitor the cable and restore power automatically once the fault is removed. Some models must be manually reset by pressing a button or cycling power at the source.

In addition to single-port injectors, several manufacturers offer multiport injectors, including 4-, 6-, or 12-port models. These models may be more economical or convenient for installations where many devices will be powered through the CAT 5 cable originating in a single wiring closet or from a single switch. They typically operate in exactly the same manner as their single-port counterparts.

Devices that are not PoE compatible can be converted to PoE with DC pickers or taps, which are sometimes called *active Ethernet splitters*. Such devices pick off the DC voltage that has been injected into the CAT 5 cable by the injector and makes it available to the equipment through the regular DC power jack.

Two types of pickers and taps are available: passive and regulated. A passive device simply takes the voltage from the CAT 5 cable and directs it to the AP for direct connection. If 48 volts DC is injected by the injector, then 48 volts DC will be produced at the output of the passive device. A regulated device takes the voltage on the CAT 5 cable and converts it to another voltage. Several standard regulated voltages are available: 12 volts DC, 6 volts DC, and 5 volts DC. This enables a wide variety of non-PoE equipment to be powered through the CAT 5 cable.

Cable Management

In a properly designed wiring system, each cable pair should be viewed as a manageable asset that can be manipulated to satisfy any user requirement. Individual cable pairs should be color coded for easy identification. Cables should be identified at both ends by a permanent tag with a serial number. This permits the individual pairs to be selected at either end with a high degree of reliability and connected to a patch panel or punch-down block as appropriate to implement moves, adds, or changes.

Patch panels and punch-down blocks should be designed to segregate different functions or circuit types to expedite the location of pairs. Data and voice connections, for example, can be terminated in different areas of the panel or block to avoid confusion. Many panels and blocks are precolored or have color-tagging capabilities, which can assist in the location of pairs.

The cabling used in wiring a telephone system or LAN depends on the requirements of the devices being used and the formal distribution plan that the telephone system and/or computer vendor provides. Formal plans, such as IBM's Cabling System, compete with similar plans advocated by other vendors. All of these plans have the common goal of establishing a wiring strategy that will support present needs and future growth. The in-house technician (or maintenance firm) should be familiar with these cabling schemes.

A number of asset management applications are available that keep track of the wiring associated with connectors, patch panels, and wiring hubs. These cable management products offer color maps and floor plans that are used to illustrate the cabling infrastructure of one or more offices, floors, and buildings. Managers can create both logical and physical views of their facilities, and even view a complete data path simply by clicking on a connection.

Some products provide complete cable topologies, showing the locations of the cabling and connections, providing views of cross-connect cabling, network diagrams by floor, and patch panels and racks (see Figure 3-19). Work orders can be generated for moving equipment or rewiring, complete with a picture of the connections. With this information, the network administrator knows where the equipment should go, what needs to be dis-

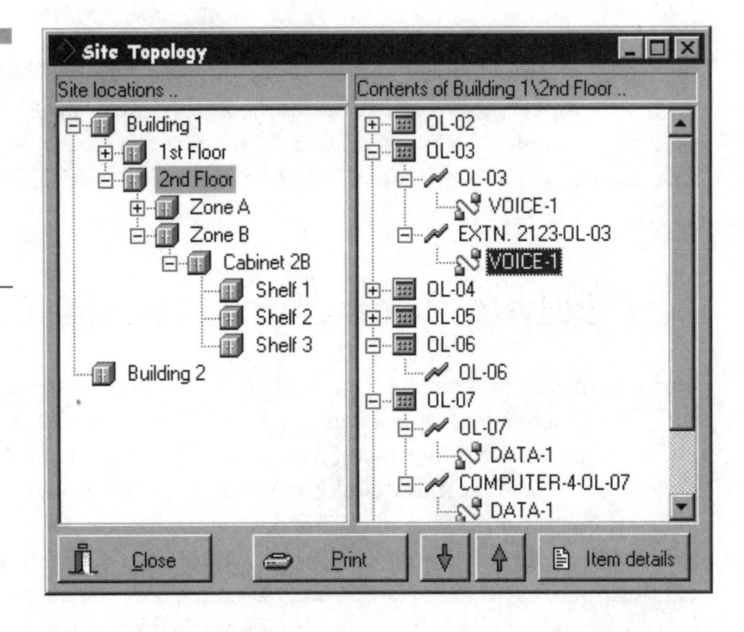

Figure 3-19 A view of site topology from Unylogix Technologies' Cable Management System

connected, and what should be reconnected. The technician can take this job description to the location and perform the changes.

Other products provide a *computer-aided design* (CAD) interface, enabling equipment locations and cable runs to be tracked through punch-down blocks, multiconductor cables, and cable trays. In addition, bill-of-materials reports can be produced for new and existing cable installations.

Cable management applications can be run as standalone systems or integrated with help desk products, hub management systems, and network management platforms. When coupled with a hub management system and help desk, a high degree of automation can be brought to bear on the problem resolution process. When the hub management system detects a media failure, the actual cable run can be extracted from the cable management application and submitted along with a trouble ticket generated by the help desk. And when the hub management system is integrated with a network management platform such as Hewlett-Packard's OpenView, all of this activity can be monitored from a single management console, which expedites problem resolution.

Determining if the existing wiring can be used to support new system installations requires the preparation of a complete wiring plan, preferably before the start of system installation. The plan acts not only as a guide to the installation process, but also as a check on the capacity and planning that can be carried out before installers appear and begin working. In the case of a hub or switch, for example, the plan should associate each computer with a complete path back to the cabling at the hub or switch through all the panels, horizontal feeds, and risers. In locations where it is not practical to install cable runs, wireless links can be implemented via APs or bridges.

Verifying Network Connectivity

Once the network is set up with cable connections from each computer to a hub, connectivity can be checked by opening Windows Explorer and viewing either Network Neighborhood (Windows 95/98/ME) or My Network Places (Windows XP). All computers that are configured to belong to the same workgroup should be listed. If they are not, make sure that firewall software is not running on the computers that are expected to show up on the network—this will hide them from view, and prevent file and printer sharing. This includes the ICF that comes with Windows XP as well as any third-party firewall software.

If you still have a problem viewing computers on the network, run the ping utility that comes with Windows. Ping is a simple test function that is used to check if a local or remote system on an IP network is currently in operation. The easiest way to use ping is to run it from the Windows command line (accessed under Run). It can also be accessed from the command prompt found under Accessories in the program list. The ping command is run using plain language domain names or IP addresses. The general command-line syntax for implementing ping is

```
ping nathan or ping 192.168.100.2
```

This will indicate whether the computer named Nathan is currently online. Ping will furnish the IP address assigned to Nathan as well (see Figure 3-20). Ping can also be used to check the operation or connectivity to a cable/DSL modem, router, web site, and DNS server—any addressable device on the internal network or the public Internet. The ping command sends one datagram per second and displays one line of output for every echo response returned. If the target computer is out of service or the connection is bad, no echo is returned. Instead, the ping function times out, indicating a problem.

With just ping, however, it cannot be determined whether the problem is with the target computer or the connection. Further investigation is

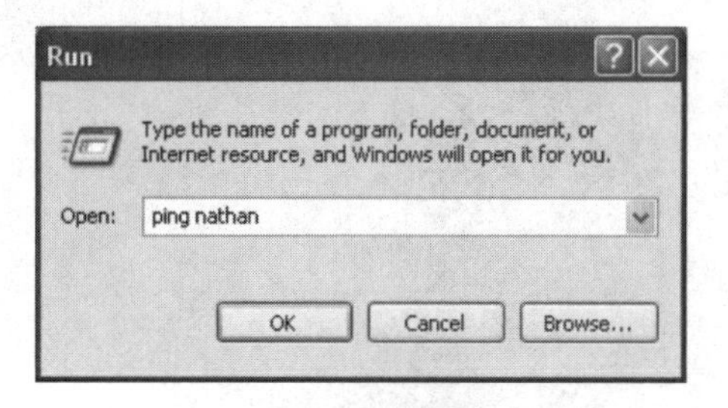

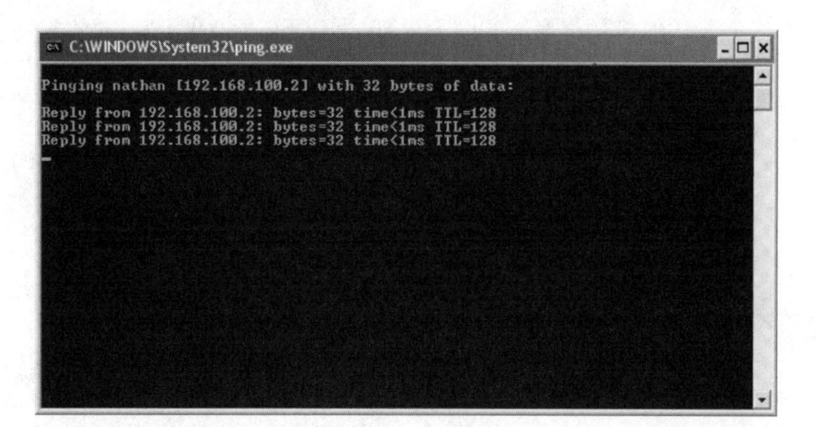

Figure 3-20 When the ping command is run from the Windows command line, packets are sent to the target computer (Nathan) and echoed back, indicating that the connection is good and that the target computer is online.

required to isolate the source of the problem. If the ping command fails, verify that the remote host IP address is correct, that the remote host is operational, and that all of the gateways (routers) between the computer and the remote host are operational. If you still have a problem running the ping command, it could be that the packet-filtering policies on routers, firewalls, or other types of security gateways might prevent ping requests from reaching their destination.

There could also be a problem with the cable or connectors somewhere between the source and target computers and the hub on the internal network. Some problems with cable include breaks, miswires, shorts, or cross pins. Faulty NICs are another possible problem. To test for these and other problems on an internal network, a portable Ethernet cable tester can be used (see Figure 3-21), especially for verifying network connections

Figure 3-21 An Ethernet cable tester with a remote unit from Cables N Mor

during the initial installation and whenever moves, adds, and changes occur. Some testers can also provide measurements of cable length to ensure they meet specifications. The test unit itself has a master and remote component into which both ends of the cable are inserted. The tester supplies the electrical power to drive signals across the cable so anomalies can be detected.

Conclusion

Whether or not a computer is connected to a peer-to-peer or client-server network, it must have an NIC and be configured for TCP/IP networking. Even if the computer will be used in a wireless environment, it is a good idea to also configure it for use on a wired LAN. In case the computer is moved to a location that does not have or is out of range of a wireless connection, a LAN connection may be used. Computers can support other types of networks as well, such as Bluetooth and *General Packet Radio Service* (GPRS). It is just a matter of disabling one type of connection and enabling another type of connection as the circumstance warrants.

CHAPTER 4

Preparing for Wireless Deployment

With *wireless local area network* (WLAN) technology becoming more secure, reliable, and affordable than ever before, enterprises of all types and sizes are starting to take advantage of the mobility and productivity benefits that WLANs offer. *Wireless Fidelity* (WiFi) products, especially, are quite easy to install. In fact, an organization might already have wireless links in operation without even knowing it. Employees are bringing their own WiFi equipment into the workplace similar to the way they brought their own PCs into the workplace in the 1980s and for the same reason: to empower themselves. A security audit could find numerous wireless *access points* (APs) that the company knows nothing about.

The simplicity and affordability of WiFi equipment, however, masks an array of critical issues. Users who are not getting the bandwidth they expect, for example, might add more APs only to see bandwidth plummet further. Another concern is that an AP could radiate signals far beyond the office, offering hackers a way into the corporate network, where they can view, steal, or damage sensitive or irreplaceable information. A telecommuter who uses an AP at home and has a *virtual private network* (VPN) link to the corporate LAN may unwittingly invite hackers into the VPN by not having the proper security safeguards in place. Consequently, a successful WLAN deployment is not as simple as plugging in an AP and handing out wireless cards; it requires a careful assessment of network and user requirements, site preparation, and a sound security strategy.

Wireless support can be justified from a number of perspectives. Employees want to bring notebook computers and *personal digital assistants* (PDAs) to conference rooms, cafeterias, other offices, and even to other buildings in order to collaborate with colleagues at convenient places and accomplish their work at locations more conducive to productivity or concentration. A person who works all day long in a windowless 8×10 office with the constant buzz of fluorescent lighting, for example, might actually be more productive if he or she could occasionally take a notebook computer outside to the company picnic table.

Sometimes it is difficult for employees to work in cubicle environments, which offer little or no privacy amid the constant din of chatter, laughter, and phone calls. Writers, programmers, and accountants especially need a quiet place to work and could be more productive if they could just move their notebook computer to an unoccupied office or conference room for long periods of silence, without having to worry about finding a data port for plugging into the corporate LAN.

Managers may actually spend very little time in their office during the course of a workday. They spend much of their time in meetings with superiors and subordinates. On many projects, they may have to visit with peers from other departments. Instead of being out of touch much of the day and spending extra hours catching up on e-mail, a wireless connection would let them carry a notebook computer wherever they go and be productive during the downtime between meetings.

Salespeople spend much of their time visiting prospects and customers during the day. A good deal of time is spent traveling by car from one location to another within a region. Instead of having to go back to the office to send files and take care of e-mail, productivity can be enhanced by having a corporate account with a local *wireless Internet service provider* (WISP). This would enable salespeople to use their time productively between customer visits. In some cases, a wireless connection might even enable salespeople to look up items such as pricing, inventory, and shipping status right in the customer's office. This enhances customer response, which also adds to the professional image of the salesperson and company.

Although wireless connections offer obvious utility in warehouses and other wide-open spaces, disagreement still exists about whether WLANs are suitable for general office environments. This issue needs to be discussed among department heads within the organization before a serious planning effort even gets off the ground. For some companies, the deployment of 802.11b WLANs is discouraged because actual bandwidth is limited to about 5 to 6 Mbps. But even at these speeds, WLANs can support a range of business applications, including *enterprise resource planning* (ERP) and the distribution of PowerPoint presentations. Large file downloads, such as *computer-aided design/computer-aided manufacturing* (CAD/CAM) drawings, may have to be forced onto the wired LAN or use 802.11a wireless links that offer up to 54 Mbps.

Although some companies may not view 802.11b as a drop-in replacement for wired infrastructure and may refuse to consider it at all, other companies limit wireless to conference rooms or other semipublic areas, relying on the wired network for daily use. Other companies have committed to deploying wireless throughout their organization, even among multisite campuses, treating it as a viable extension of the corporate network. At these companies, workers have come to expect the convenience of wireless access. If a group gathers for a meeting in one location that does not have wireless coverage, they are apt move to another location that has it.

Needs Assessment

A place to start when determining the need for wireless connectivity throughout the organization is to conduct a formal needs assessment, which can be facilitated by circulating a questionnaire that can be e-mailed or otherwise distributed to all employees. The questionnaire can be used to elicit important information from employees, which can be used to determine what kind of equipment to buy, the placement of APs and bridges, the type of security that must be provided, and the kind of administrative tools and *information technology* (IT) expertise that may have to be added. Among the essential elements that should be addressed by the questionnaire are

- What kind of computer and applications does the employee currently use?
- How mobile is the employee throughout the workday and at what locations?
- What workgroups and departments does the employee interface with on a regular basis?
- Does the employee currently have a conveniently located data port?
- What kinds of files are typically transferred internally and what are the sizes in kilobytes or megabytes?
- Are the files considered fairly routine or are they "company sensitive" and not for public disclosure?
- Is the employee satisfied with the throughput rate of the current wired LAN?
- How often does the employee access the Internet and how long are the sessions?
- Does the employee work away from the corporate office? If so, how much time is spent away and where?
- Does the employee travel frequently, spending a lot of time at airports and in hotels?

The decision to go wireless may be made on a case-by-case basis, weighing business needs at a given corporate site against criteria such as security and costs. The expertise at each site to handle such evaluations, however, may vary widely. The IT staff may handle the process with help from equipment vendors or consultants. The IT department should formalize a process to help users and workgroups evaluate the need for wireless before they submit their requests for wireless connectivity.

Sometimes opportunities will arise that will prompt the consideration of a WLAN, such as when

- Retrofitting an existing network
- Adding or expanding a secondary network
- Adding, moving, or changing employees to new locations

- Acquiring temporary office space to accommodate growth
- Hiring temporary office staff
- Organizing workspace for a project workgroup
- Increasing mobility for part or all of the workforce
- Enhancing network performance
- Expanding business operations

Performance Baselining

An essential task in preparing for WLAN deployment is to take a snapshot of current LAN performance. This is accomplished with performance baselining, a procedure that enables IT managers to understand the behavior of a properly functioning network so that deviations can help identify the cause of problems that may occur in the future. The only way to know a network's normal behavior is to analyze it while it is operating properly. Later, technicians and network managers can compare data from the properly functioning network with data gathered after conditions have begun to deteriorate. This comparison often points to the steps that lead to a corrective solution.

With a snapshot of problem-free LAN performance, any deviation after the wireless components are intalled should be interpreted as being caused by those components. That way, IT managers and technicians will know where to look to find and correct the problem. If the problem is not immediately apparent, the wireless components—APs and bridges—can be disconnected from the network to see if the network returns to baseline performance. If the network returns to baseline performance, the problem has been conclusively narrowed down to one or more wireless components. More details about troubleshooting WLANs are provided in Chapter 10, "Selecting a Wireless Internet Service Provider."

The first step in baselining performance is to gather the appropriate information from the properly functioning network. Much of this information may already exist—it is just a matter of finding it.

Topological Map

For example, many enterprise management systems have the capability of automatically discovering devices on the network and creating a topological map. This kind of information is necessary for knowing which components exist on the network and how they interact physically and logically. For *wide area networks* (WANs), this may mean locations, descriptions, and cable plant maps for equipment such as routers, bridges, and network access devices.

In addition, information on transmission media, physical interfaces, line speeds, encoding methods and framing, and APs to service provider equipment should be assembled. Although it is not always practical to map individual workstations in the LAN portions of the WAN, or to know exactly what routing occurs in the WAN cloud, knowing the general topology of the WAN can be useful in tracking down problems later.

WAN and LAN Protocols

To fully understand how a network behaves, it is necessary to know which protocols are in use. Later, during the troubleshooting process, the presence of unexpected protocols may provide clues as to why network devices appear to be malfunctioning or why data transfer errors or failures are occurring.

Logs

Some network problems begin to occur after new devices or applications are installed. The addition of new devices, for example, can cause network problems that have a ripple effect throughout the network. A new end-user device with a duplicate *Internet Protocol* (IP) address, for instance, could make it impossible for other network elements to communicate, or a badly configured router added to the network could produce congestion and connection problems. Other problems can occur when new data communication services are enabled or existing topologies and configurations are changed. A log of these activities can help pinpoint the causes of network difficulty. In addition, previous network trouble—and its resolution—is sometimes recorded, which can also lead to faster problem identification and resolution.

Statistics

Often, previously gathered data can provide a valuable context for newly created baselines. Previously assembled baselines may also contain event and error statistics and examples of decoded traffic based on the network location or time of day. These logs may have been gathered over long periods, yielding valuable information about the history of network performance.

Usage Patterns

A profile of users and their typical usage patterns can also speed fault isolation. This entails having several types of information, including what kind of LAN traffic is carried over the WAN.

LAN Traffic on the WAN

With knowledge of what kind of LAN traffic to expect on the WAN, technicians and network managers will have a better idea of the analysis that might have to be performed later. In addition, knowing how LAN frames might be handled at end stations can help troubleshooters make a distinction between WAN problems and end-station processing problems.

Traffic Content

Knowing the WAN traffic type (voice, data, video, and so on) can help troubleshooters estimate when network traffic is most likely to be heavy, what level of transmit errors can be tolerated, and whether it makes sense to even use a protocol analyzer. For example, an analyzer may incorrectly report errored frames and corrupt data when attempting to process voice or video traffic based on data communication protocols.

Peak Usage

Knowing when large data transfers will occur, such as scheduled file backups between LANs connected across the WAN, can help network managers predict and plan for network slowdowns and failures. It can also help technicians schedule repairs so that WAN performance is minimally impacted. Some of this information can be obtained from interviews with network administrators or key users. Other times, it must be gathered with network analysis tools.

Hard Stats and Decodes

After gathering information on topology, devices, protocols, and typical users of the WAN, hard statistics and examples of decoded network traffic should be gathered. Getting comprehensive baseline data may entail gathering it at regular intervals at numerous points throughout the network.

Statistics Logs

To understand usage trends and normal error levels over time, a statistics log is created. Many protocol analyzers let technicians specify the period over which this kind of data is logged, the interval between log entries, and the type of statistics to log. The log file can be exported to a spreadsheet or other application program for offline analysis.

Frame or Packet Data

To see details about typical WAN traffic, frame or packet data can be collected and saved to a file for later examination. Data collection can be done at specific periods during the day or week to find differences between peak and off-peak usage. Saved network traffic also provides insight into device configurations for use later during routine upgrades or repairs.

Targeted Statistics

Using configurable traffic filters and counters, selected blocks of data or statistics based on specific network events can be captured, which might include error count thresholds, specific frame types, and in-channel alarms. A comprehensive collection of such data provides a benchmark for comparison if the network begins to malfunction. New protocols on the network, unexpected line and channel utilization levels, and increases in normal errors and in-channel alarms can be isolated according to physical link location, helping narrow the search for the problem.

If network performance and reliability problems occur, the information gathered during baselining can be used to help identify the nature and source of the problem through comparison analysis and historical trends.

Comparison Analysis

Baseline information is compared with current information to see network changes. For example, to isolate failing devices or connections, the number of errors recorded during baselining is compared to the current number of errors that occur over a similar time interval.

Historical Trends

Current network problems can result from subtly changing conditions that are detected only after examining a series of baselines gathered over time. For example, congestion problems may become apparent only as new users are added to a particular part of the network. Examining historical trends can help isolate these situations.

In conclusion, performance baselining provides a profile of normal network behavior, making it easier for technicians and network managers to identify deviations so the appropriate corrective action can be taken. This snapshot of the current network can also be used as the input data for subsequent performance modeling. For example, network administrators and operations managers can use the baseline data to conduct what-if scenarios to assess the impact of proposed changes, including the addition of wireless bridges and APs to the network.

A variety of other changes also can be evaluated, such as adding routers, WAN bandwidth, or application workloads, and relocating user sites. During analysis, performance thresholds can be customized to highlight network conditions of interest. These capabilities enable IT managers to plan and quantify the benefits of feature migrations, such as different routing protocols, and to make more accurate and cost-effective decisions regarding the location and timing of upgrades.

Another reason for examining the wired network is to ascertain the traffic patterns and bandwidth demand typical of the user population. After all, the wireless network will be an extension of the existing network. Based on this review, IT managers can start to estimate what throughput, coverage, and security will be needed for a given set of applications. These considerations, in turn, guide planners through the process of evaluating different wireless interface cards, APs, and bridges. One thing often overlooked is that wireless equipment must be cabled to the LAN. Ethernet jacks may have to be installed so APs can be attached to the corporate LAN, and electrical power outlets may be needed for the APs, though some vendors offer the option of powering the access devices over *Category 5* (CAT 5) cable.

Asset Management

Asset management entails the proper accounting of various hardware assets within the organization, including wireless assets such as WiFi network adapters, APs, and bridges—any device that has an IP address. Other types of assets that can be tracked include software, network lines, and internal cabling.

Without a thorough understanding an organization's assets, it will not be possible to accurately plan departmental budgets or allocate costs. The failure to account for and manage assets such as desktop hardware and software, in-house cabling, and network lines has other ramifications as well. It can lead to cost overruns on projects, leave the door open to employee theft (asset shrinkage), and lead to the misuse or abuse of the network. The lack of controls can expose the company to financial penalties for copyright infringement, as when employees copy software or the organization allows more concurrent usage than the vendor's license permits.

Similar concerns apply to the wireless environment. Once a wireless network is implemented, the assets must be secured and managed to leverage the benefits they promise. But if no asset management system is in place, it can be difficult to control the network. For example, APs are quite inexpensive, so individual employees or workgroups could easily purchase them on their own and set up an unauthorized wireless network. Being vulnerable in their default configurations, unauthorized APs often create security holes in the network. If an unauthorized AP is added to the network, it must be discovered and the situation must be remedied. Without asset management tools, reaping the rewards of a wireless environment can be frustrating and very costly.

Types of Assets

Organizations must track several types of assets. As noted, these fall into the general categories of hardware, software, network, and cable assets. With the right asset management tool, these assets can be automatically discovered on the network and tracked through their life cycles. Typically, a scan is conducted by the asset management application running on a collection server, which retrieves configuration details from all the client devices on the network.

All files that are transferred between the collection server and the client are compressed to conserve bandwidth. Some asset management products offer block-by-block check pointing so that if a client is interrupted for any reason, the inventory process picks up where it left off down to the file block level during the next update. If a client is disconnected from the network during a scan, the inventory will continue running and the results will be uploaded when the connection is reestablished.

Hardware Assets Hardware inventory starts with identifying the major kinds of systems in the distributed computing environment—from the bridges, APs, and servers all the way down to the desktop, notebook, and even PDAs, as well as any components they may have, including the *central processing unit* (CPU), memory, boards, and storage. The asset management utilities that come with servers generally scan connected devices for this kind of information.

Most asset management products provide the following basic hardware information:

- **CPU** Model and vendor
- **Memory** Type (extended or expanded) and amount (in kilogytes, megabytes, or gigabytes)
- **Hard disk** Amount and percentage of disk space used and available, volume number, and directories
- **Ports** In use and available
- **Addresses** *Media Access Control* (MAC) address and IP address

Most hardware identification is based on the premise that if a driver is loaded, then the associated hardware must be present. However, many of these drivers go unused and are not removed, resulting in inaccurate inventory. This situation is remedied by industry standards, such as the *Desktop Management Interface* (DMI) and *Plug and Play* (PnP).

Hardware inventories can be updated automatically on a scheduled basis—daily, weekly, or monthly. The hardware inventory typically includes the physical location of the unit, owner (workgroup or department), and name of the user. Other information may include vendor contact information and the unit's maintenance history. All this information is manually entered and updated. When all resources are scanned, an inventory report may be printed.

In addition to providing inventory and maintenance management, some products provide procurement management as well. They maintain a catalog of authorized products from preferred suppliers, as well as list and discount prices. They track all purchase requests, orders, and deliveries. With some products, even the receipt of new equipment can be automated, with the system collecting information from scans of asset tags and bar codes. Warranty information can also be added.

Still other asset management packages accommodate additional information for financial reporting, such as

- **Cost** Purchase price of the unit and add-in components
- **Payment schedule** Principal and interest
- **Depreciation** One-time expense or multiyear schedule
- **Taxes** Local, state, and federal (as applicable)
- **Lease** Terms and conditions
- **Charge-back** Cost charged against the budgets of departments, workgroups, users, or projects

This kind of information is manually entered and updated in the asset management database. Depending on the product, this information can be exported to spreadsheets and other financial applications, and be used for budget monitoring, expense planning, and tax preparation. Although wireless equipment such as network adapters and APs are very inexpensive, they are worth tracking in these terms because hundreds or thousands of such devices may be distributed across multiple locations. This adds up to a significant capital investment.

Software Assets Another technology asset that must be tracked is client software, including firmware. Not only can software tracking (also called applications metering) reduce support costs, but it can also protect the company from litigation resulting from claims of copyright infringement, such as when users copy and distribute software on the network in violation of the vendor's license agreement.

Asset management products that support software tracking automatically discover what software is being used on each system on the network by scanning local hard drives and file servers for all installed software. They do this by looking for the names of all executable files and arranging them in alphabetical order. They determine how many copies of the executable files are installed and look into them to provide the product name and publisher. Files that cannot be identified absolutely are listed as found but flagged as unidentified. Once the file is eventually identified, the administrator can fill in the missing information.

The accumulated asset information can be used to build a software distribution list. The administrator can then automatically install future upgrades on each workstation appearing on the distribution list. With wireless network adapters, this capability can be used to update firmware to accommodate new features, changes in standards, and security patches. The updates are distributed over the network from a central management location in the same way that client software upgrades are handled.

The administrator can also monitor the usage status of all software on the network to enforce license compliance. If several copies of an application are not being used, they can be made available to other users. If all copies of an application are in use, a queue is started and the application is made available on a first-come, first-served basis. This kind of control may qualify the organization for discounts on software. A growing number of software vendors will not sell network licenses unless the customer has a metering system in place. If a metering system is not in place, the organization may be forced to pay a higher price for unlimited software usage.

Network Assets The kinds of network assets that must be monitored, controlled, and accounted for in inventory include repeaters, bridges, routers, gateways, hubs, and switches—any wired or wireless device that is used to implement the network. An enterprise management system may offer asset management as a native function or permit the integration of a third-party application that offers this capability.

The entire chassis of a hub or switch, for example, can be viewed via the network management system, which shows the types of cards that are inserted into each slot. With a zoom feature, any card can be isolated and a representation of the ports, *light-emitting diodes* (LEDs), and configuration switches can be displayed at the management console. A list of devices attached to any given port, including wireless APs and bridges, can also be displayed and/or printed. Some of the other views available from the management system include the following:

- *Configuration view* organizes the configuration values for a device and its model, including the device location, model name, firmware version, IP address, and security string.
- *Resource view* is a special-purpose view for endpoint devices, showing where the endpoint device accesses its application resources, such as primary and secondary print servers, the e-mail server, and file server.
- *Cable walk view* illustrates the connections that exist along a segment of cable and the devices connected to each segment, including any wireless APs.
- *Diagnostic view* organizes diagnostic and troubleshooting information for a device, including errors, collisions, events, and alarms.

- *Performance view* displays performance statistics, including load, hard and soft errors, and frame traffic.
- *Port performance views* summarize port- or board-specific performance statistics for each individual port or board.
- *Application view* organizes device application information, including device IP and *Internet Control Message Protocol* (ICMP) statistics.
- *Assigns view* enables a specific technician to be assigned to devices owned by network users. For example, a technician with *radio frequency* (RF) expertise would be assigned to handle problems with wireless APs and bridges.

The management system may also provide a method for documenting the equipment and cable plant inventory through a third-party cable asset management system. This and other third-party applications are typically integrated with *application programming interfaces* (APIs).

Cable Assets Even with a wireless infrastructure in place, the need for cabling will always exist. Therefore, among the assets that also must be managed is the cabling that connects all the devices on the network, including coaxial cable (thick and thin), twisted pair (shielded and unshielded), and optical fiber (single-mode and multimode).

A number of specialized applications are available that keep track of the wiring associated with connectors, patch panels, cross-connects, and wiring hubs. They use a graphical library of system components to display a network. Clicking on any system component brings up the entire data path, with all its connection points. These cable management products offer color maps and floor plans that are used to illustrate the cabling infrastructure in one or more buildings. A zoom feature can isolate backbone cables within a building, on a floor, or within an office.

Some cable asset management products can generate work orders for moving equipment or rewiring. Managers can create both logical and physical views of their facilities, and even view a complete data path simply by clicking a connection. Some cable asset management products automatically validate the cabling architecture by checking the continuity of the data paths and the type of network for every wire. In addition, a complete picture of the connections can be generated and printed out. With this information, network administrators and technicians know where new equipment should go, what needs to be disconnected, and what should be reconnected. Some cable asset management products can even calculate network load statistics to facilitate proactive management and troubleshooting.

Like other types of asset management applications, cable management applications can be run as standalone systems or can be integrated with help desk products, hub management systems, and major enterprise management platforms offered by companies such as Computer Associates, Hewlett-Packard, and IBM/Tivoli.

Web-Based Asset Management

For organizations that understand the value of asset management but do not have the time for it, inventory services are available that provide instant, accurate inventory over the Web. Tally Systems, for example, offers WebCensus as a fast, centralized, low-maintenance way to audit enterprise PCs without having to install or manage anything. The tool securely and efficiently audits PCs and returns hardware and software inventory results in minutes (see Figure 4-1).

IT managers collect inventory data by sending users an e-mail with an embedded link to the service provider's web site. When users click the link, an agent is installed onto the PC, performs an inventory, and sends it to a database on the service provider's site before uninstalling itself. All data is encrypted during transport. IT administrators then log onto the site and run reports against the database. The reports can be used to determine such things as how many copies of Microsoft Office are deployed and which versions. The reports can be saved as an Excel spreadsheet. Aside from tracking inventory, the data can be used to plan for operating system and application upgrades. However, the necessity for end users to initiate the inventory collection is a weak link in the process. If the link in the e-mail is ignored, the inventory process is not initiated.

If everyone cooperates, the advantage of a subscription-based inventory service is that it enables IT managers to offload management costs and pay only for the services they use, rather than invest in a software product and incur its associated installation and maintenance costs. As a hosted service, it completely eliminates the need for installing, configuring, and updating complex software. This reduces stress on IT staff, protects the information system infrastructure, and eliminates the need for in-house upgrades. Web-based inventory services are available in 1-, 3-, and 12-month subscriptions and are priced from $3 to $15 per PC on the basis of the length of subscription and type of service.

Although this type of service is not yet available for wireless equipment on the corporate network, this may change in the near future as more telecommunications carriers and national ISPs get into the market for

Figure 4-1 Workstation Details is one of about 20 reports offered by Tally Systems' WebCensus, an online inventory service.

Details for Workstation BBBouchard

User	Login	Group	
Bruce Benjamin Bouchard	BBBouchard	Finance	
MAC Address	**IP Address**	**Serial Number**	
00C04F8B3677	192.169.4.654	SN16173	
Total Disk Space (MB)	**Free Disk Space (MB)**	**Total Memory (MB)**	**Last Inventory**
7683	3438	64	12/14/1999 12:43:29 PM

System

Dell OptiPlex G1 266MTbr+

Operating System

Microsoft Windows 95 4.00

Hardware Components

Category	SubCategory	Manufacturer	Product
			System Board
		Novell	NetWare Shell Driver 3.26-0
			Memory Module
			Memory Module
			Memory Module
BIOS	BIOS	Phoenix	ROM BIOS
CD/DVD	CD-ROM		CD-ROM Drive
Diskette	Diskette		Diskette Drive
Hard Drive	Hard Drive	Quantum	Fireball ST2.1A
Keyboard	101/102 key		101/102 keyboard
LAN Adapter	LAN Adapter	Intel	EtherExpress PRO/100B
Logical Drive	Logical Drive		FAT-16 Partition - Big DOS
Logical Drive	Logical Drive		FAT-16 Partition - Big DOS
Logical Drive	Logical Drive		FAT-16 Partition - Big DOS
Monitor	VGA	Dell	D825TM
Monitor	VGA	Gateway	CrystalScan 1572DG
Mouse	Serial Mouse	IBM compatible	PS/2 Mouse
Parallel Port	Parallel Port		Parallel Ports
Processor	Pentium II	Intel	Pentium II
Serial Port	Serial Port		Serial Ports
Video Adapter	VGA	ATI	3D RAGE IIC Controller

Software Components

Category	SubCategory	Manufacturer	Product
Asset Mgmt	PC Inventory	Tally Systems	NetCensus Win32 Collector Unknown
Comm Software	Integrated	Microsoft	Exchange Client for Win32 5.0
Comm Software	Internet Tools	Microsoft	NetMeeting for Win32 2.1
Comm Software	Internet Tools	Netscape Communications	Netscape Communicator for Win32 4.5
Comm Software	Fax	Cracchiolo and Feder	RightFAX Client for Windows 5.0
Database	DB Managers	Sybase	Sybase SQL Anywhere for Win32 5.0
Database	DB Managers	Microsoft	Access 97 8.0 SR-2
Games	Misc	Microsoft	Minesweeper 95
Games	Misc	Microsoft	Hearts Network for Win32 95
Games	Misc	Microsoft	FreeCell 95
Games	Misc	Microsoft	Solitaire 95
Graphics	Presentation	Microsoft	PowerPoint 97 8.0 SR-2
Integrated	Integrated	Microsoft	Office 97 Taskbar 8.0
Multimedia	Integrated	Macromedia	ShockWave 7 7.0
Spreadsheet	Spreadsheets	Microsoft	Excel 97 8.0 SR-2
Utility	PIM/Cont. Mgr.	Microsoft	Schedule+ for Win95 7.5
Utility	PIM/Cont. Mgr.	Microsoft	Outlook 98 8.5
Utility	Data Compress.	Nico Mak Computing	WinZip for Win32 7.0 SR-1
Word Processor	Word Processors	Microsoft	Word 97 8.0 SR-2

Open in Excel

wireless Internet access service. Wireless infrastructure management services, including security and asset management, offer ways for these service providers to differentiate themselves in an increasingly competitive marketplace.

In summary, asset management can have several approaches. Organizations can buy one or more software packages, use an integrated approach available with some help desk or network management systems, or outsource the asset management task to a systems integrator or computer vendor. In addition to containing the cost of technology acquisitions and reining in hidden costs, such asset management can improve help desk operations, enhance security, assist with technology migrations, minimize asset shrinkage, and provide essential information for planning network expansion and upgrades.

Security

Although security often comes up later in the deployment phase, it should be considered earlier because of the inherent vulnerabilities of wireless technology. Wireless connections inherently are not secure; data is broadcast through the air and is hard to contain. The original encryption scheme for 802.11 WLANs, called *Wired Equivalent Privacy* (WEP), is known to have several weaknesses that are exploited by hackers to decode the captured data.

In deciding whether to deploy wireless technology, it must be recognized early on that considerable resources may have to be devoted to security. The IT department should endeavor to understand what the vulnerabilities are and what can be done to mitigate them. It may be a good idea for the IT department to hire someone with experience in wireless technology and the special security issues involved. That person should be able to recommend and implement precautions to minimize risks and stay current with new and emerging risks.

It would be a mistake for a company to ignore security by claiming that its data has no value to anyone else. Hackers are not always interested in stealing data; they may want to do other things such as destroy data, shut down the network, or get into the Accounts Payable system to create a fictitious company to which checks would be issued for bogus services. Some hackers might be interested in using a company's computers to launch distributed *denial of service* (DoS) attacks to disrupt another organization's network and information systems.

Depending on the specific requirements of the organization, security can range from turning on the basic WEP encryption and changing all default settings on wireless equipment to full-blown authentication and encryption via VPNs tied into *Remote Access Dial-In User Service* (RADIUS) servers. A good approach to security would be to treat the WLAN as if it were the public Internet and take the appropriate countermeasures. This involves using hacker tools to assess vulnerability and plug potential breach points. A firewall should be considered for placement between the wireless links and the wired LAN.

Although vendors often tout VPNs as the ultimate security solution, this comes with a number of trade-offs. Administration becomes more complicated, requiring the distribution of VPN client software and upgrades to hundreds or thousands of devices. When administrators made changes or updates, they previously had to send them to one device at a time. Policy management capabilities greatly facilitate the configuration of firewalls and installation of digital certificates onto VPNs. Although administrators can dynamically see and build VPNs from a central spot with policy management, and as easy and convenient as all this sounds, it still consumes staff time and requires an investment in the right tools.

A related issue is that most employees with wireless notebooks do not realize their wireless cards remain active, even if they are not using the VPN. Thus, it is possible for a hacker to use that active link to access a notebook computer and infect it with a virus or other malicious code, which is transmitted to the corporate network via the VPN when the employee logs on. Worse, the hacker can use the VPN link to access any corporate resource and steal information or wreak other kinds of havoc to cripple corporate operations. A hacker can also create a personal backdoor to the corporate network, so he or she can enter anytime from any location, regardless of whether the VPN connection is available. All these issues can be effectively addressed, but require an investment of resources.

A hidden expense associated with the deployment of WLANs is employee education, particularly on security matters. Employees must be made aware of their responsibilities in using wireless technology. Above all, they must not be allowed to install unauthorized APs within the organization because it may compromise the entire network. They also should not be allowed to have hacker tools installed on their computers. This can be readily determined with periodic scans of client software by the asset management application that runs on a server.

To aid understanding and retention, the vulnerabilities of wireless connections should be clearly spelled out to employees, perhaps with the aid of diagrams. A security policy should be put in writing and each employee should be required to sign it. A willful action that results in a security breach should result in a warning that becomes part of the personnel file or termination of employment, depending on the seriousness of the offense and whether it is a first or repeat offense.

Site Survey

The actual design of the wireless infrastructure—how many APs and bridges are needed to ensure adequate coverage—draws on how the issues previously discussed were resolved, plus all the data accumulated by the asset management system and the statistics collected about the problem-free performance of the existing wired LAN. The next step toward deployment of the wireless infrastructure is the site survey, which seeks to identify appropriate locations for the equipment.

The placement of equipment will hinge on factors such as the type of materials used in building construction and furnishings, the number of users in a given area, whether that number changes, and the throughput those users need. The larger the coverage area and the more demanding the applications, the more complicated the deployment becomes. Although a small business can arrive at the proper placement of one or two APs on an intuitive basis, larger companies that need hundreds of APs must use asset management and baseline data as well as design tools to arrive at the optimal solution.

Regardless of the size of the organization, it is advisable to physically examine each location before installing an AP. During this inspection, the building layout or site map should be marked up to show all coverage areas. Starting with the building outline, mark the outside areas that need coverage. Identify possible obstructions for RF signals such as freezers, coolers, X-ray rooms, elevators, and microwave ovens in employee breakrooms and cafeterias. Because metal reflects RF signals, collections of office equipment such as metal bookshelves and cabinets also constitute potential RF problem areas. These possible obstructions must be marked as well.

If due care is not exercised in the placement of APs and bridges by taking into account all the variables that can affect performance and coverage, one risks compromising network security, paying for more equipment than is really needed, not giving employees the throughput they require for their applications, and wasting money on wireless technology that does not fulfill organizational needs.

A site survey is also essential for dealing with the problem of overlapping channels. The 802.11b APs, for example, have a maximum of three nonoverlapping channels for users. Too many APs, arbitrarily placed, will overlap these channels, causing users to experience a serious drop in performance because of channel contention. Proper channel configuration enables three APs to be in close proximity to each other and still provide users with the maximum available bandwidth.

The 802.11a APs, on the other hand, have eight indoor channels and four outdoor channels. This means that more APs can be packed into the same area to support more users at higher bandwidth, but only if proper attention is given to channel configuration.

The same tools that are used to plan and document wired networks can be used for wireless networks. Among the popular off-the-shelf tools that can be used for this purpose are AutoCAD and Microsoft Visio. Of the two, Visio is easier to use. Its graphical representation of complex projects enables more people to understand and participate in the planning process without having to spend a lot of time learning to use it. Shapes of wireless equipment can be selected from a device library and inserted into the workspace. For each device, attributes can be added to further describe the device by parameters such as vendor, make and model, serial number, MAC and IP address, and location.

Tools are now available that go far beyond mere drawing and documentation capabilities, offering more intelligence to aid in the proper design of 802.11 WLANs. Wireless Valley Communications, for example, offers a product called LANPlanner, a toolset for predicting, measuring, and optimizing WLANs. It also comes with an asset management capability. The following tools available with LANPlanner are:

- *Predictor* provides a full-color visual representation of the performance of the wireless network, including vital information such as the *Received Signal Strength Indicator* (RSSI), *Signal-to-Interference Ratio* (SIR), and *Signal-to-Noise Ratio* (SNR).
- *LANFielder* verifies and records network performance statistics directly from the *Institute of Electrical and Electronics Engineers* (IEEE) 802.11a or 802.11b wireless network adapters. Some of the network performance statistics that can be collected include received signal strength, packet throughput, error rate, latency, and jitter.
- *Optimatic* uses field measurement data taken with LANFielder to optimize wireless system predictions. The optimization process derives

the average error between measured and predicted RF coverage throughout multistory and/or multibuilding environments. The optimized model can be archived for use in a future project in similar surroundings to take advantage of previous measurement knowledge.

LANPlanner keeps records of all technical, fiscal, and maintenance details for each project. Since most indoor and campus networks involve hidden wiring and APs that are not easy to find and can be lost over time, LANPlanner's graphical models let IT managers visualize where all the infrastructure equipment is located, directly on an electronic blueprint, so IT staff can find any network component.

After the site survey is completed, all findings should be summarized in a report for the installation team, including the building diagram depicting AP placement and cell structures, antenna choice, configuration parameters, power requirements, possible interference sources, and digital photographs of each location.

Design Process

A designer can take a top-down or bottom-up approach to building the wireless network. In the former, the designer starts by using the appropriate tools to sketch out the overall network; subsequent drawings add increasing levels of detail until every aspect of the network is eventually fleshed out. The bottom-up approach might start with placing the wired LAN on a specific floor of a specific building and adding APs with subsequent drawings linked to create the overall infrastructure.

For a multisite wireless network, the designer may start with an aerial photo or a scaled or scanned drawing of a campus or urban core. Using a layered approach, the designer can place APs, bridges, or repeaters on top of that image. The resulting coverage, capacity, and cost of the wireless infrastructure can then be more easily visualized.

Once the wireless systems are actually installed and properly aligned, taking into consideration the signal's Fresnel Zone as well as its line of sight, other tools can be used to test and adjust signal strength, simulate traffic between them, and test for data throughput and an application response. Other tools can be used to check for signal leakage beyond the intended coverage area with the idea of revealing potential security vulnerabilities.

A comprehensive site survey helps determine the most precise placement of the APs. In addition to diagramming and documenting the site, surveying involves measuring the strength of the signal. Although signal strength can reveal whether the signal is strong enough to be received, SNR measurements are used to compare the signal to the noise floor. If noise in the band is high enough, it can cause reception problems, even if a strong signal is made from the AP. Because signal strength alone is not sufficient, using both SNR measurements and the packet retry count (the number of times packets have to be retransmitted for successful reception) is an accurate way to validate a coverage area.

Ideally, packet retry count should be below 10 percent in all areas. It can be used to determine the edge of RF data reception. The signal may be strong in certain areas, but because of the noise floor or multipath interference, the signal cannot be decoded by the receiving device, and the packet retry count will increase. Without an SNR reading, however, it will not be known whether packet retries are increasing because the devices are out of range, the noise is too high, or signal strength is too low.

The site survey also must determine if RF interference exists. An 802.11b WLAN uses the 2.4 GHz band, while an 802.11a WLAN uses the 5 GHz band. Both of these are shared, unlicensed bands. Because they use shared spectrum, neglecting to take into consideration interference created by microwave ovens, cordless phones, satellite systems, and other wireless devices such as RF lighting systems and neighboring WLANs can seriously affect performance.

Equipment Considerations

A common mistake is to use one brand of interface card and AP for the initial network design and then switch to a different brand in the final deployment. This can lead to surprises stemming from different RF propagation characteristics, which leads to dead spots and lower bandwidth. It is best to research available products before doing any RF measurements and base the purchase decision on such criteria as vendor stability, product quality, features, management capabilities, and upgrade potential. That way, the IT department will not feel pressured to change its commitment with every new product announcement from alternative sources and possibly upset a carefully designed wireless infrastructure.

For example, a company might like 802.11a for its 54 Mbps throughput versus the stated throughput of 11 Mbps for 802.11b. Many companies that had plans to install 802.11b equipment and start pilots for 802.11a now have no reason for doing so. This is because 54 Mbps will become available for 802.11b using the 2.4 GHz RF band.

One popular configuration that is emerging consists of using 802.11b to create blanket site coverage at a maximum usable bandwidth of 4 to 6 Mbps, with an eye to using 802.11a products to create higher-bandwidth hot spots for power users or high-bandwidth applications. Some vendors, including Agere, Cisco, and Intel, offer APs with two card slots so users can add 802.11a when needed. By using a dual-band AP, organizations can preserve investments in existing 11 Mbps WiFi equipment while supporting 54 Mbps WiFi5 hot spots for special purposes.

This is not to say that no good reasons exist for standardizing the entire organization on 802.11a. An 802.11a WLAN should be considered when

- The savings gained through higher system performance is more important than initial system cost.
- An urgent need exists for bandwidth and speed to handle large graphics, audio, data, and video files.
- A network capacity greater than what an 802.11b WLAN can currently deliver is needed.
- The WLAN must match the speed of wired networks.
- There will be many users per AP.
- The current wired network is to be replaced by a wireless one.

A different set of circumstances may warrant consideration of an 802.11b WLAN, despite its current bandwidth limitation. An 802.11b WLAN should be considered when

- Initial system cost is the primary concern.
- Data rates up to 11 Mbps are sufficient for current applications and the current number of users.
- The goal is global implementation.
- An existing 802.11b WLAN needs to be expanded.
- Handheld PCs need WLAN access.
- A small number of users per AP exist.
- A small number of users need to be added to the existing wired LAN.

Because data rates affect the range, selecting data rates during the network design stage is very important. APs offer clients multiple data rates for the wireless link. For 802.11b links, the range is from 1 to 11 Mbps in four increments, while for 802.11a links, the range is from 6 to 54 Mbps in seven increments. The client cards automatically switch to the fastest possible rate supported by the AP. Each data rate has a unique area of coverage or cell size: the higher the data rate, the smaller the coverage area or cell size. The minimum data rate for any given cell must be determined early in the design stage. Selecting only the highest data rate will require a greater number of APs to cover a given area. This means care must be taken to develop a compromise between the required aggregate data rate and overall system cost.

Numerous vendors offer wireless APs and network adapters. At the low end, vendors provide inexpensive 802.11b devices aimed at home users who have little or no technical expertise. Consequently, such APS and network adapters require minimal configuration and offer limited opportunity for customization. At the high end, vendors offer premium-priced, enterprise-class devices. Enterprise APs, for example, may have metal covers, special seals for harsh environments, a choice of antenna, network management software, an advanced web-based user interface for administrators, a range of specialized software, and many proprietary features. These vendor-exclusive features might include support for higher bandwidths using proprietary techniques that the IEEE standard does not sanction.

Although these vendor-exclusive features can frustrate interoperability, a number of other factors can also render systems incompatible. One is software: Drivers available for one brand of network cards may not work with another brand of AP. Another factor is that a given product may not come with the drivers needed to get it operating properly. In some locations, a specialized antenna may be needed to "shape" and direct the radio transmission. In that case, the AP must be able to accept an external antenna.

Deployment

If the planned deployment of wireless technology is large, it is recommended that the equipment be unpacked at a staging area and configured there before sending items out to each location. At this time, the network names and identification databases can be established, network and security settings can be loaded into the APs, and all the equipment can be tested for proper operation. Since the site survey will identify the optimal location of APs, it is just a matter of making sure the cables are installed and power outlets are available at those locations.

If outdoor units are required to extend coverage, these need to be properly enclosed and grounded. Following the manufacturer's instructions will make such installations go smoothly and result in trouble-free operation well into the future. If the outdoor units are located near a recreational area, extra protection may be needed to prevent accidental damage from errant softballs, basketballs, or volleyballs.

Expect ongoing adjustments. Employees might pile equipment around an AP, for example, causing network performance to drop. New shelving, new walls, and shifts in inventory—any change in the operating environment—can affect network throughput.

When an AP stops working for whatever reason, fixing it can create other problems. To get the device working again, power can be shut off and turned back on again, forcing the equipment back into proper operation. If this does not work, the device might have to be reset, which returns it to the default configuration. The reset may clear out the security settings, leaving the AP vulnerable to hackers. Nobody will know for sure unless the security settings are specifically checked after the reset.

Another deployment consideration is how roaming will be facilitated. After all, users may want to move from office to office, floor to floor, or even building to building with their portable computers and PDAs. Consequently, users may need to move across subnets. Some operating systems, such as Windows XP and Windows 2000, support automatic *Dynamic Host Configuration Protocol* (DHCP) release/renew to obtain the IP address for the new subnet. Certain IP applications such as VPNs, however, will fail when this feature is enabled. This issue can be solved by deploying a flat network design for the WLAN, where all APs in a roaming area are on the same segment.

Many large companies use several flat networks. They determine where users will typically roam and try to segment the wireless network based on coverage areas with a minimum number of users roaming between them. If no coverage exists outdoors, users will lose IP connectivity as they move between buildings, so a good segmentation plan entails establishing one subnet per building.

With all equipment installed at their proper locations, the wireless network should be thoroughly tested at all levels, checking security policies, throughput, and coverage before turning it over to live traffic.

Conclusion

As easy as it is to get a wireless link up and running, doing it properly takes as much up-front planning and more ongoing diligence than a traditional wired network. If the requirements of the wireless network can be clearly defined, a deployment plan can be developed that will not only save money, but will also increase employee autonomy, productivity, mobility, and job satisfaction. Knowing users and their application requirements will help define coverage areas that do not compromise security by allowing signals to range beyond intended coverage areas. Surveying from the outside in ensures that coverage areas do not extend beyond the building.

CHAPTER 5

Access Points and Clients

Several components on the *Wireless Fidelity* (WiFi) network need to be configured to interoperate securely in the corporate environment; among them are *access points* (AP) and clients. The following discussion draws upon a variety of WiFi products to provide a sense of what is required for proper configuration and the reasons behind the different setup choices; however, it is not intended to be used as a replacement for the more detailed documentation that comes with specific products. This discussion starts with the configuration of APs and continues to the configuration of clients, so the clients have something to connect to after setup.

Access Point Setup

A wireless AP usually comes preconfigured from the factory with default settings. Because the same equipment can be used in the home and office environment, the default settings will need to be changed so the AP can operate more efficiently and securely as part of the enterprise network. The device's configuration utility is used to change the default settings and customize the unit for operation on the corporate network. The configuration utility can be accessed through a web browser or through the AP's console port using a terminal emulator. If certain configuration parameters are used, they must be set the same for the AP and its associated clients.

The AP comes with a *Media Access Control* (MAC) address and a default *Internet Protocol* (IP) address. The default IP address is used to access the web-enabled configuration utility. Later, the default IP address can be changed so that it falls within the same range as other IP addresses used on the corporate network. The network administrator can choose an IP address or have one assigned to the AP by another device on the network, such as a router or server, via the *Dynamic Host Configuration Protocol* (DHCP).

To configure an AP through the web-enabled utility, the unit must first be connected to a computer using a length of standard *Category 5* (CAT 5) cable. After this connection is made, the user opens the web browser and types in the AP's default IP address on the command line, which might look something like http://192.168.0.50/. Upon pressing Enter, the configuration utility stored in the AP comes up on the computer screen with various fields that need to be filled in. The fields are described in the following sections, but because every configuration utility is different, they are not presented in any particular order.

Username and Password

The first task is to enter a username and password. This will prevent unauthorized access to the configuration utility. Although this is not foolproof, it usually thwarts casual users. After clicking OK at the bottom of the screen, the Statistics window comes up, showing the MAC address of the devices on the network. Because the AP is directly cabled to a computer for first-time configuration, only the MAC address of the AP will be listed. As clients are configured and associated with the AP, their MAC address will be listed as well (see Figure 5-1).

Next, the Configuration window is opened where the network administrator can change the default settings of the AP. The default settings enable the AP to be installed and put into operation right out of the box. Although this is intended to make things easier for consumers, it is risky to leave the default settings in place when the device is intended for installation on a corporate network. This is because the default settings leave the device vulnerable to hackers who can find the wireless link and use it to gain access to the network and all the sensitive information stored there.

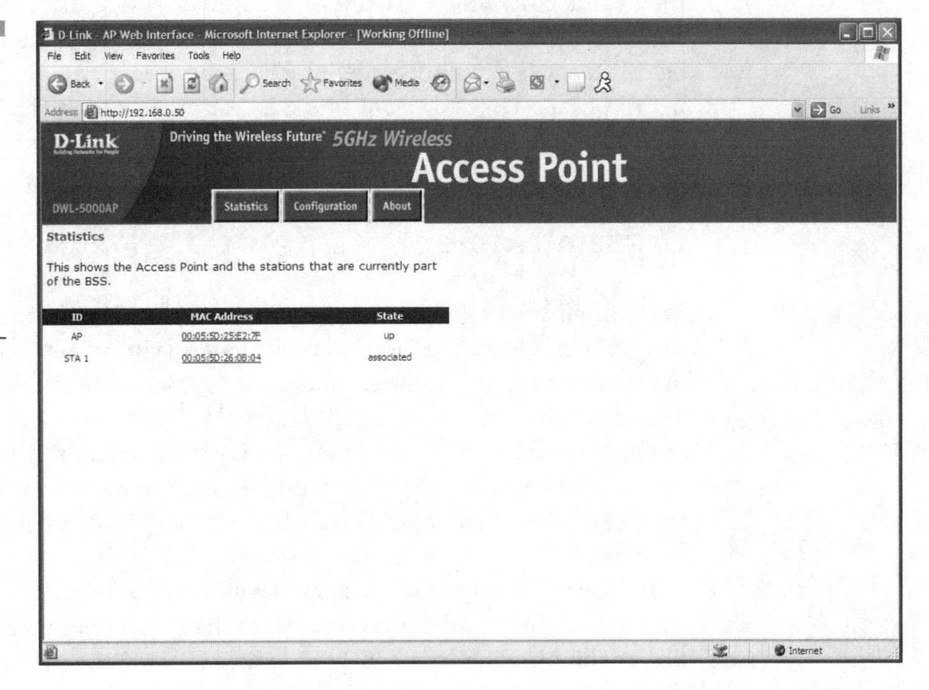

Figure 5-1 The Statistics window of the 5 GHz DWL-5000 Access Point from D-Link, showing the MAC address of the access point and one associated client

Service Set Identifier (SSID)

One of the default settings that should be changed without hesitation is the *service set identifier* (SSID), which identifies the network. The SSID can be any alphanumeric entry that is 2 to 32 characters long, such as Cisco_Aironet_AP1500. It is similar in concept to the workgroup name for a peer-to-peer network and is intended as a simple security feature. Client devices configured with the same SSID can get onto the network through the AP they are associated with. The use of SSIDs also makes it easy to segment wireless networks, so multiple workgroups can coexist in the same wireless space without interfering with each other. A related feature, SSID broadcasting, advertises the presence of the AP so client devices with the same SSID can easily connect at start-up.

On the other hand, these features also make it easier for hackers to find the network. The default SSIDs are even published in vendor literature and are well known to the hacker community, rendering their value for security useless, even if they are changed. Many network administrators simply turn them off to avoid becoming easy prey for hackers. With SSID broadcasting turned off, client devices merely send out a probe asking if the SSID they want is available. If the SSID is available, link-up is accomplished automatically. Because such transactions occur infrequently rather than being broadcasted continuously, there is much less chance that hackers will notice SSIDs. Therefore, not much is lost by disabling the SSID broadcast feature except a slight delay in getting onto the network.

System Name

In addition to the MAC address, the network administrator can assign a system name to the AP. This is typically descriptive text, such as Accounting Department or Conference Room A-1, which is helpful in identifying the location of the AP. This is an optional field that has no bearing on the operation of the AP. In fact, many network administrators refrain from entering any location-specific information into this field because it can potentially tip off hackers about where to aim their equipment to snoop for sensitive information.

In mixed vendor environments, the system name can be the model of the AP, such as AirPro AP1200. Sometimes the system name can include a portion of the MAC address, such as AirPro AP1200-xxxxxx, where xxxxxx is the last six digits of the AP MAC address. The system name field, although not required, is often used for entering helpful administrative information.

Radio Channel

The AP usually comes with a default radio channel. It does not need to be changed unless another WiFi device nearby happens to be using the same channel. The same channel cannot be used by APs that are within range of each other without causing interference. The use of different channels also eliminates the bandwidth contention that occurs when two APs with overlapping coverage are configured with the same channel. When this happens, the 802.11 wireless Ethernet *carrier sense multiple access/collision avoidance* (CSMA/CA) mechanisms ensure that users in both coverage areas can access the network. However, instead of providing two separate 11 Mbps channels with an aggregate bandwidth of 22 Mbps, the two APs provide only one 11 Mbps channel.

The 802.11b standard defines 14 *center frequency channels* (see Figure 5-2), 11 of which are authorized for use in the United States. To prevent signal overlaps with several adjacent channel frequencies, in practice only 3 channels can be used—channel 1 at 2.412 GHz, channel 6 at 2.437 GHz, and channel 11 at 2.462 GHz. These three channels, when laid out correctly, can accommodate large installations with many APs and clients (see Figure 5-3).

The configuration utilities of 2.4 GHz systems typically list all 11 channels. When the network administrator sets up a network of multiple APs, care must be taken to select channels that are separated far enough in frequency range that they do not overlap in case some of the APs are in range of each other. Likewise, the configuration utilities of 5 GHz systems typically list all eight channels, and the same frequency spacing precaution must be taken when selecting channels in multisystem environments.

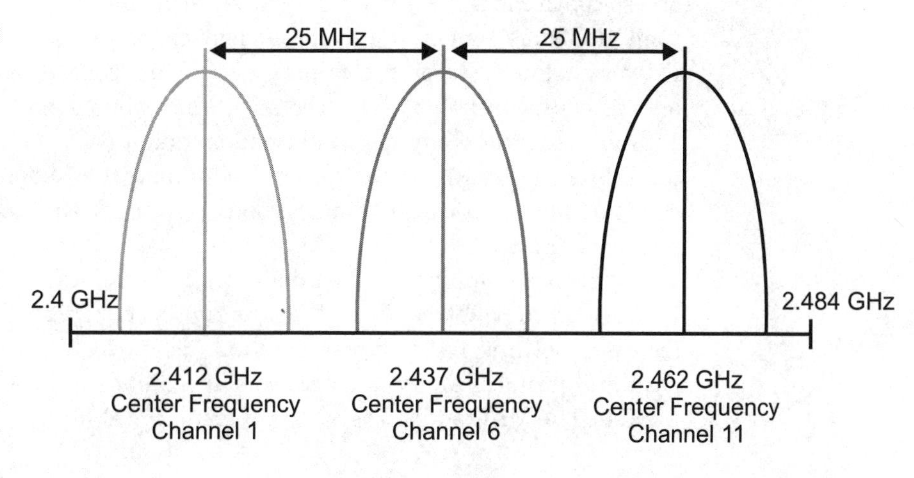

Figure 5-2 Standard 802.11b specifies 14 center frequency channels. Of the 11 channels authorized for use in the United States, only 3 can be used to provide nonoverlapping coverage.

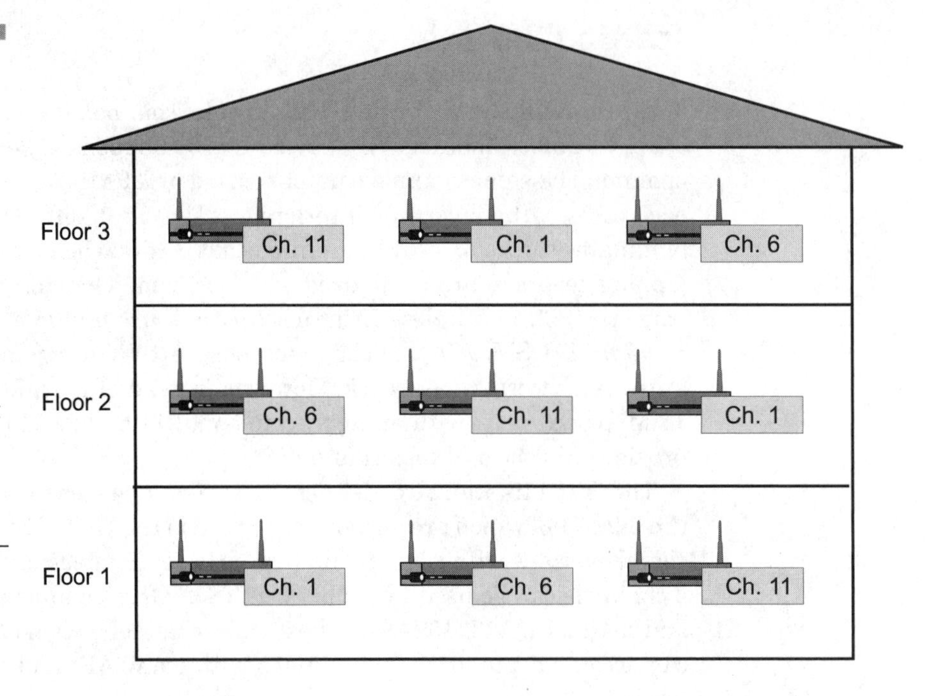

Figure 5-3 A sample frequency topology using channels 1, 6, and 11 on access points distributed on three floors of an office building. Note that access points on the same channel are never placed next to each other, which minimizes the chance of interference.

MAC Filtering

To increase security, APs permit the MAC address of every authorized client device to be entered into a table called the *access control list* (ACL). Devices listed in the table with their MAC address are allowed onto the network; those not listed are denied access to the network. Although this method of access control is fine for small networks with one or two APs, it is cumbersome and time consuming to implement on larger networks because the network administrator must maintain an up-to-date list of these MAC addresses and reprogram all the APs whenever people join or leave the organization and equipment is issued or taken back. The effectiveness of MAC filtering hinges on the network administrator keeping up with table changes, but it can also unintentionally shut off the network to highly mobile users.

But even assuming that the table is well maintained, it is relatively easy for hackers to penetrate the AP through *packet sniffing*. A packet-sniffing device is nothing more than a notebook computer equipped with a WiFi card and protocol analysis software that enables a hacker to decode the packets. The MAC address is a component of the Ethernet frame data. Once

MAC addresses have been sniffed, an intruder can reconfigure the MAC address of their own hardware to match an allowed address or spoof one to make it look like a legitimate MAC address, thereby gaining admittance to the network. Depending on the network adapter, this change can be accomplished through the Windows Control Panel.

You do not even have to be a skilled hacker to overcome SSIDs and MAC filtering—an employee leaving the company is likely to know all the necessary SSIDs and MAC addresses to get back on the network. All he or she has to do is sit in the parking lot and gain access to the network and/or sniff the traffic radiating from the building. SSIDs and MAC filtering should never be relied upon by themselves for security.

Authentication

The network administrator can choose between two types of authentication: open and shared key. With open authentication (the default setting), any client can begin a conversation with an AP without being challenged. Because all clients are considered authenticated, this provides no security whatsoever. With shared key authentication, client adapters can communicate only with APs that have the same secret key. The AP sends a known unencrypted *challenge packet* to the client adapter, which encrypts the packet and sends it back to the AP. The AP attempts to decrypt the encrypted packet and sends an authentication response packet indicating the success or failure of the decryption back to the client adapter. If the packet is successfully encrypted/decrypted, the user is considered authenticated.

Encryption

The standard for WiFi encryption is called *Wired Equivalent Privacy* (WEP), which is available in different iterations based on the level of security desired. WEP uses either 40- or 128-bit keys in hexadecimal format that must match between client device and AP. A 40/64-bit WEP key is only 10 hexadecimal digits, whereas 128-bit WEP uses 26 hexadecimal digits. Some vendors refer to 40-bit encryption as 64-bit encryption. They are both the same; the difference stems from a lack of standardization in terminology. Once the shared key is entered and applied, it will no longer be visible in the configuration field. Asterisks will appear in place of the hexadecimal digits.

The shared key is secret and only known to the client device and AP. The secret key is used to encrypt packets before they are transmitted and decrypt them after reception. An integrity check is applied to ensure that packets are not modified in transit. Many network administrators have chosen not to enable WEP for several reasons. One reason is that all clients who access the system need to have the key. In a large environment where potentially hundreds of clients have the key, secrecy becomes nonexistent. WEP has been described as giving everyone in the company the same password and never changing it. Newer protocols are emerging that offer solutions for this problem, but they have not matured yet (see Chapter 7, "Wireless Security").

Another reason for rejecting WEP is that a degree of protocol overhead and latency is associated with using WEP encryption. Also, some incompatibility exists between vendor implementations and many IT organizations want to preserve the flexibility to choose equipment from multiple sources. Finally, the WEP encryption mechanism can easily be cracked with widely available hacker tools such as AirSnort, which is specifically designed to recover encryption keys. It operates by passively monitoring transmissions and computing the encryption key when enough packets have been gathered. AirSnort requires approximately 5 to 10 million encrypted packets to be collected. Once enough packets are sampled, AirSnort can come up with the right encryption password in under a second.

If you want to configure encryption anyway, you can select the type of encryption you want to implement usually from a pull-down list. Depending on the type of product, the following choices might be available:

- None
- 64 bits (10 hex digits)
- 128 bits (26 hex digits)
- 152 bits (32 hex digits)

After selecting the type of encryption, enter the key in the appropriate field using hexadecimal format. Hexadecimal digits consist of the numbers 0 through 9 and the letters *A* through *F*. For 64-bit encryption, you must enter exactly 10 hexadecimal digits into the encryption key field; for 128-bit encryption, you must enter exactly 26 hexadecimal digits. Some 5 GHz WiFi products offer 152-bit encryption, which requires that you enter exactly 32 hexadecimal digits in the encryption key field. An example of a 10-digit hexadecimal encryption key is 67BF47ACDE. The encryption key you devise for the AP must also be configured for each client device that will be affiliated with it.

Network Role

Some APs can be configured for a specific role in the wireless network. When this feature is available, the configuration utility enables the administrator to select an appropriate role from a drop-down list. The menu may contain the following options:

- **Root AP** This is a wireless transceiver that connects an Ethernet network with wireless client stations. This setting is used if the AP will be connected to the wired *local area network* (LAN).
- **Repeater AP** This is an AP that transfers data between a client and another AP. This setting is used for APs not connected to the wired LAN. As a repeater, the AP boosts signal strength to provide coverage over a greater distance.
- **Site survey client** This is a client device with a wireless connection to an AP that is used for diagnostics, such as when the administrator needs to test the AP by having it communicate with another AP.
- **Radio network optimization** This feature enables the network administrator to either keep preconfigured settings or customize them as follows:
    - **Throughput** This setting maximizes the data volume handled by the AP, but might considerably reduce the AP's range.
    - **Range** This setting maximizes the AP's range, but might considerably reduce its throughput.
    - **Custom** This setting enables network administrators to enter their own settings in an attempt to more precisely optimize the performance of the AP.

Turbo Mode

Depending on manufacturer, some APs have a proprietary turbo mode feature. When enabled, turbo mode enables traffic to burst higher than the 802.11 wireless standards allow. For example, a 2.4 GHz system (802.11b) normally permits up to 11 Mbps, but in turbo mode, it can burst up to 22 Mbps. A 5 GHz system (802.11a) normally permits up to 54 Mbps, but in turbo mode, it can burst up to 72 Mbps or higher.

If the AP is configured for turbo mode, the client devices must also be configured that way for the feature to work. This means having equipment from the same manufacturer at both ends of the wireless link. (A follow-on standard, 802.11g, will enable 2.4 GHz systems to pass data at up to 54 Mbps, but the standard is not final at this writing. It will be backward compatible with existing 11 Mbps systems.) In turbo mode, the range of the wireless link will be considerably less.

IP Address

As noted, APs come from the factory with a default IP address, which is used to access the configuration utility from a web browser. This address can be left as is or changed so that it falls within the range of other IP addresses used on the corporate network.

Configuration Server Protocol

From a drop-down list, the network administrator can select a configuration server protocol, which specifies the method of IP address assignment for the AP. The choices displayed in the drop-down list include the following:

- **None** This option is selected when the network does not have an automatic system for IP address assignment, as when a small business relies on a local *Internet service provider* (ISP) for one or more static IP addresses.
- **BOOTP** This option refers to the *Bootstrap Protocol*, in which a server on the network assigns the IP addresses to the APs based on their MAC addresses.
- **DHCP** This option refers to the DHCP, which is implemented from another device on the network such as a router, gateway, or server, which assigns the IP addresses to the APs. The IP addresses are leased for predetermined periods of time.

Subnet Mask

Entering an IP subnet mask identifies the subnetwork so the AP's IP address can be recognized on the LAN. The default subnet mask is 255.255.255.0, which suffices for most installations where Class C addresses are used; otherwise, the subnet mask can be changed.

Default Gateway Address

This is the IP address of the Internet gateway—the device that actually sends and receives data between a user's network and the Internet. This device is usually a router. It could also be another type of device that contains router functionality, such as a cable or *Digital Subscriber Line* (DSL) modem, or an *integrated access device* (IAD) that combines voice, data, and Internet access over the same digital line.

Administrative Access

The AP's configuration utility enables a new user to be added to a list of people authorized to view and make changes to the AP's management system. From a list, the network administrator can select from among the following capabilities to assign the new user:

- **Write** The user can change system settings.
- **Identity** The user can change the AP's identity settings (IP address and SSID).
- **Firmware** The user can update the AP's firmware.
- **Admin** The user can view most system screens. To allow the user to view all system screens and make changes to the system, the write capability is selected.

Using a Terminal Emulator

In addition to being able to configure APs with a web browser through a CAT 5 cable connection, terminal emulators can be used to set initial AP settings through the unit's console port. Microsoft's HyperTerminal is an example of a terminal emulator, but others can be used as well. The *command-line interface* (CLI) provides the same functionality as a browser interface. Commands are entered in text form or selected from a list of choices.

Typically, the network administrator uses the serial interface when the AP is initially being configured with basic settings. The serial port is easily accessible on the back of the unit. When the AP is initially configured and mounted in an elevated operating location—such as a wall, pole, or rooftop—the browser interface is more convenient to use when making AP configuration changes.

With the serial cable properly installed between a computer and the AP, the terminal emulator program is opened and the network administrator enters the following settings:

- Bits per second (baud rate): 9600
- Data bits: 8
- Parity: No parity
- Stop bits: 1
- Flow control: Xon/Xoff

Next, the manufacturer's procedure is followed to display the home page of the AP. This procedure may differ for first-time configurations and reconfigurations. This tab is used to set the number of Telnet sessions, password, and other values. By default, the Telnet port number is 23. Access can be controlled by setting a password. The maximum number of concurrent Telnet sessions can also be set in the configuration utility. The number of seconds to wait for a login after connection can be set, after which the AP connection closes automatically. The session can also be set to time out after a specified number of seconds if no activity occurs.

Client Setup

Client devices—desktop computers, notebooks, and *personal digital assistants* (PDAs)—must be equipped with a WiFi network adapter, which comes in several form factors, as described in Chapter 3, "Setting Up the Network." Once the drivers and configuration utility for the network adapter are installed, the utility can be opened and the information can be entered into the appropriate fields in much the same way as was done with the AP.

System Parameters

The client configuration utility enables the network administrator to set system, *radio frequency* (RF), infrastructure, and security parameters that prepare the client adapter for use on a wireless network (see Figure 5-4).

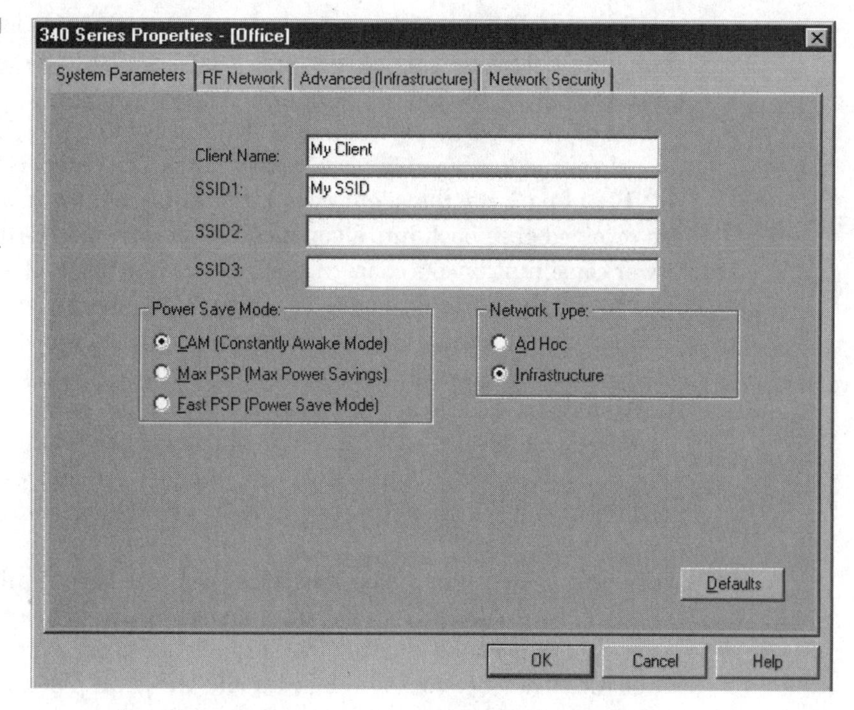

Figure 5-4 Configuration utility for setting system parameters for Cisco 340 Series client adapters

Client Name This is a unique plain language name for the workstation. It enables the administrator to determine what devices are connected to the AP without having to memorize every MAC address. This name is included in the AP's list of connected devices. The field allows up to 16 characters.

SSID The SSID identifies the specific wireless network to access. Up to 32 characters (case sensitive) can be used for the SSID, which must be the same as that of the AP the client will associate with. If this parameter is left blank, the client adapter can associate with any AP on the network that is configured to broadcast SSIDs. If the AP is not configured to allow broadcast SSIDs, the client adapter will not be able to access the network. Optional SSIDs may be entered in additional fields to identify other wireless networks, enabling the user to roam across different networks without having to reconfigure the client adapter.

Power Setting This field sets the client adapter's power consumption. You may select from up to three modes, the names of which differ by manufacturer. The awake mode keeps the client adapter powered up continuously so there is little lag in message response time. Leaving the client in continuous awake mode consumes the most power, but offers the highest throughput. This setting is recommended for desktop and notebook computers and portable devices when AC power is used.

Two power-save modes are available. One switches between awake mode and sleep mode, depending on network traffic. When retrieving a large number of packets, the client goes into awake mode and after the packets are received goes back into sleep mode. This setting is recommended when power consumption is a concern, such as when a notebook's battery is being used to supply power and no AC source is immediately available.

The other power-save mode causes the AP to buffer incoming messages for the client adapter, which wakes up periodically and polls the AP to see if any stored messages are waiting for it. The adapter can request each message and then go back to sleep. This mode conserves the most power, but offers the lowest throughput. It is recommended for devices for which power consumption is a critical concern, such as small battery-powered PDAs.

Network Type Two types of wireless networks are available: ad hoc and infrastructure. Also known as peer to peer, ad hoc indicates that the wireless network consists of a few wireless devices that are not connected to a wired Ethernet network through an AP. For example, an ad hoc network could be set up between computers in a conference room so users can share information during a meeting. Infrastructure indicates that the wireless network is connected to a wired Ethernet network through an AP.

RF Network

With the client configuration utility, the network administrator can set parameters that control how and when the client adapter transmits and receives data (see Figure 5-5).

Data Rate In this field, the rate at which the client adapter handles packets to or from APs (in infrastructure mode) or other clients (in ad hoc mode) is set. For infrastructure mode, automatic rate selection is recommended; the client uses the 11 Mbps data rate whenever possible, but adjusts to lower rates when necessary. Selecting a specific data rate is recommended for ad hoc mode. Selecting 1 Mbps offers the greatest range but the lowest throughput, whereas selecting 2 Mbps offers less range but greater throughput. The 5.5 Mbps option offers less range but still greater throughput, whereas selecting 11 Mbps offers the greatest throughput but

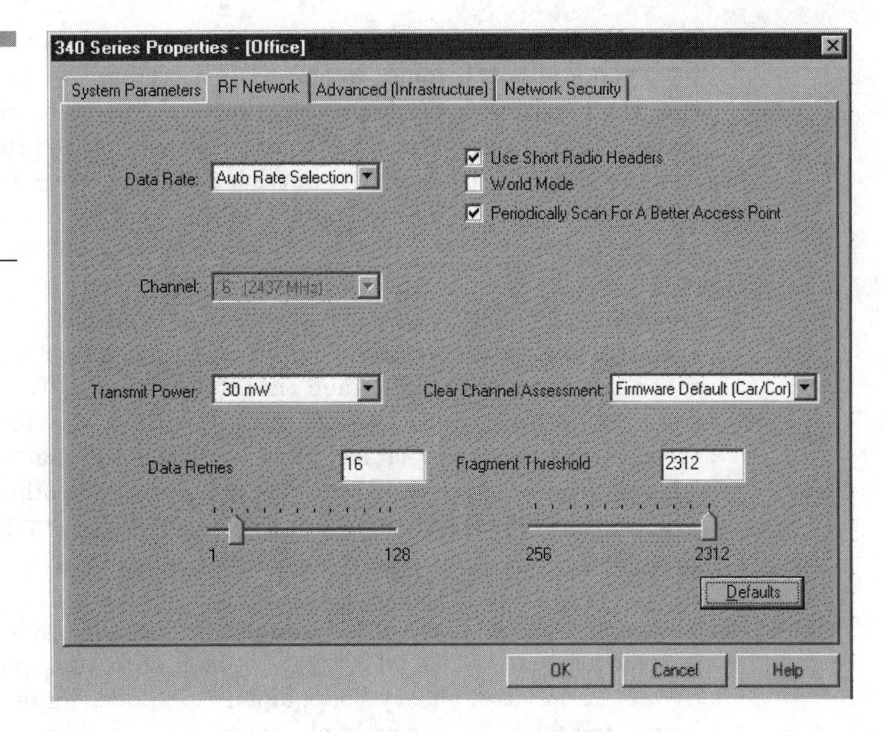

Figure 5-5 The RF Network screen of Cisco's configuration utility for 340 Series client adapters

the lowest range. The client adapter's data rate must match the setting of the AP. If it does not, the client will not be able to associate with the AP.

Short Radio Headers Some clients can be configured to use short radio headers, but only if the AP is similarly configured. Short radio headers improve throughput performance, whereas long radio headers ensure compatibility with clients and APs that do not support short radio headers. If any clients associated with an AP are using long headers, however, then all clients in that cell must also use long headers.

World Mode Available only in infrastructure mode, this setting enables the client to adopt the maximum transmit power level and the frequency range of the AP to which it is associated, provided the AP is also configured for this mode. This feature is intended for users who travel between countries and want their client adapters to associate with APs in different regulatory domains. When enabled, the client adapter is limited to the maximum transmit power level allowed by the country's regulatory agency.

Periodic Scan This setting enables the client to look for a better AP if its signal strength becomes weak and then switch associations if it finds one. The resulting signal handoff works much like a cell phone and base station. As the user drives away from a base station and the signal weakens, the signal is handed off to another base station where the signal is stronger. The process is transparent to the user, but the handoff allows uninterrupted communication as the user's location changes.

Channel This setting specifies the frequency the client adapter will use as the channel for communications with the AP. In infrastructure mode, the channel is set automatically and cannot be changed. The client adapter scans the entire spectrum, selects the best AP to associate with, and uses the same frequency as that AP. In ad hoc mode, the channel of the client adapter must be set to match the channel used by the other clients in the wireless network. The channels conform to the IEEE 802.11 standard for each regulatory domain (see Table 5-1).

Transmit Power This setting defines the power level at which the client adapter transmits. This value must not be higher than that allowed by a country's regulatory agency. For equipment intended for use in the United States, the following power levels, expressed in milliwatts, can be used: 100, 50, 30, 20, 5, or 1 mW. Reducing the transmit power level conserves battery power but decreases the radio range; conversely, increasing the transmit power increases the radio range but consumes battery power faster. In the United States, WiFi products may not exceed the power level of 100 mW.

Clear Channel Assessment (CCA) This setting specifies the method the client adapter will use to determine whether the channel is clear prior to the transmission of data. If the channel is busy with traffic from another device, the client will hold back the transmission of data until the channel becomes clear. Since data is bursty, the wait time is very short—often the user is not aware of any transmission delay. However, as more clients become active with the AP, the delay of waiting for the channel to become clear can become quite noticeable to users.

The client adapter can use several methods to determine whether a channel is clear. With the *energy-detect* (ED) method, the channel is considered busy upon the detection of any energy above the ED threshold. The energy level exceeds this threshold when data is being transmitted over the channel; after all, the transmission of 1s and 0s over any medium is simply

Table 5-1

The channel identifiers, channel center frequencies, and regulatory domains for the 2.4 GHz band

Channel Identifier	Frequency	Regulatory Domains: Americas*	Europe**	Israel	China	Japan
1	2,412 MHz	X	X	—	X	X
2	2,417 MHz	X	X	—	X	X
3	2,422 MHz	X	X	X	X	X
4	2,427 MHz	X	X	X	X	X
5	2,432 MHz	X	X	X	X	X
6	2,437 MHz	X	X	X	X	X
7	2,442 MHz	X	X	X	X	X
8	2,447 MHz	X	X	X	X	X
9	2,452 MHz	X	X	X	X	X
10	2,457 MHz	X	X	—	X	X
11	2,462 MHz	X	X	—	X	X
12	2,467 MHz	—	X	—	—	X
13	2,472 MHz	—	X	—	—	X
14	2,484 MHz	—	—	—	—	X

*Mexico is included in the Americas regulatory domain, but channels 1 through 8 are for indoor use only, whereas channels 9 through 11 can be used indoors and outdoors.

**France is included in the European regulatory domain, but only channels 10 through 13 can be used in that country.

the result of passing energy in a controlled way from one location (client) to another (AP).

Another method of detecting a clear channel is by *carrier/correlation* (Car/Cor). The channel is reported as busy upon the detection of a *direct sequence spread spectrum* (DSSS) signal, which may be above or below the ED threshold. A hybrid method of clear channel detection that reports a channel as busy upon the detection of a DSSS signal or any energy above the ED threshold. Finally, another method is based on a default *clear channel assessment* (CCA) value stored in the client adapter's firmware. The CCA default value stored in the firmware of a PCI Card, for example, might be Car/Cor, whereas the default value for the firmware of a Mini PCI Card might be ED.

Data Retries This setting defines the number of times a packet will be resent if the initial transmission is unsuccessful—in other words, the number of times a transmission will be retried if an *acknowledgment* (Ack) message is not returned by the destination device. The range for this setting is 1 to 128, with a default value of 16. If the network protocol performs its own data retries, a smaller value is entered so notification of a bad packet will be sent up the protocol stack quickly, enabling the application to retransmit the packet if necessary. If this process does not happen successfully within the stipulated number of data retries, the bad packet dies, never having been retransmitted successfully.

Fragment Threshold This setting defines the threshold above which a large data packet will be split up or fragmented. If one of those fragmented packets experiences interference during transmission, only that specific packet would need to be resent. The range is 256 to 2,312. Throughput is generally lower for fragmented packets because the fixed packet overhead consumes a higher portion of the RF bandwidth.

Infrastructure

The Infrastructure screen of the client configuration tool enables the network administrator to set parameters that control how the client adapter operates within an infrastructure network, which is simply a wireless segment connecting clients to an AP.

Antenna Mode Two antenna modes can be set for the client adapter: one for receive and one for transmit. Both are configured the same with a choice of three antenna options: diversity (both), primary, and secondary. The choice is determined in large part by the type of client adapter card. If a PC Card for a notebook computer or PDA is being configured, for example, the PC Card's built-in antenna operates best when used in diversity mode. In this mode, the card will use the better signal from its two antennas.

Another type of adapter looks like a PC Card, but it does come with a built-in antenna. This type of card is known as an LM Card and enables the user to snap on an antenna through the card's external connector. If a snap-on antenna is used, diversity mode is recommended, but primary mode can also be used. A PCI client adapter for a desktop or tower computer is configured to use only the primary antenna. The mini PCI client adapter, which can be used with one or two antennas, operates best in diversity mode.

Specify AP This setting enables the network administrator to specify the MAC addresses of APs that the client adapter is allowed to associate with. If the specified APs are not found or the client adapter roams out of range, the adapter may associate with another AP whose MAC address appears in the list. The network administrator can choose to not specify the MAC addresses of alternative APs. This will enable the client to associate with any available AP during roaming. In fact, not specifying an AP actually facilitates the roaming process, since no processing delay occurs for a MAC address lookup.

Request to Send (RTS) Threshold The IEEE 802.11b standard specifies an optional *Request to Send/Clear to Send* (RTS/CTS) protocol. This four-way handshake protocol reduces the probability of data collisions at the AP. When a sending client wants to transmit data, it first sends an RTS message and waits for the AP to reply with a CTS message (see Figure 5-6). Because all stations in the network can hear the AP, the CTS causes them to delay any intended transmissions, enabling the sending client to transmit data and receive a packet ACK.

This setting in the client configuration tool specifies the size of the data packet that the low-level RF protocol issues to an RTS packet. The range of this parameter is 0 to 2,312. A small value causes RTS packets to be sent more often and, when this occurs, more of the available bandwidth is consumed and the throughput of other network packets is reduced. On the upside, however, the system is able to recover faster from interference or collisions, which may be caused from a high multipath environment characterized by obstructions or metallic surfaces.

RTS Retry Limit This setting specifies the number of times the client adapter will resend an RTS message if it does not receive a CTS message from the AP in response to a previously sent RTS packet. The range of this parameter is 1 to 128. Using a larger value decreases the available bandwidth whenever interference occurs, but makes the system more immune to interference and collisions, which may be caused from a high multipath environment.

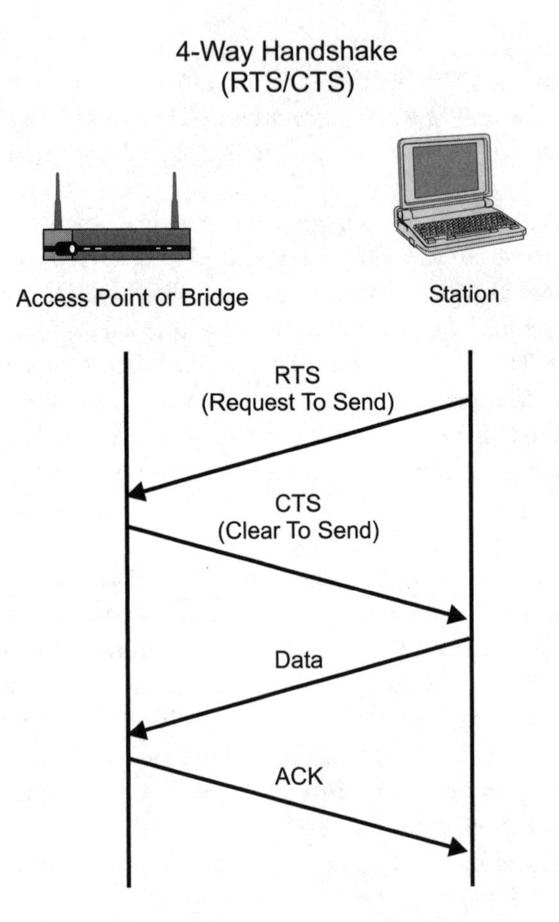

Figure 5-6 This four-way handshake occurs between the sending client and the access point. Its purpose is to minimize the chance of data collisions at the access point when multiple clients have data to send.

Ad Hoc

The Ad Hoc screen of the client configuration tool (see Figure 5-7) enables the network administrator to set parameters that control how the client adapter operates within an ad hoc network, which is simply a wireless link between clients without the use of an AP. Many of the parameters listed in the configuration utility's Ad Hoc screen are the same as those for the Infrastructure screen, but with the addition of settings for wake duration and beacon period.

Wake Duration This setting specifies the amount of time following a beacon that the client adapter stays awake to receive announcement traffic indication messages, which are sent to the adapter to keep it awake until the next beacon. This is a power-conservation feature, so the setting

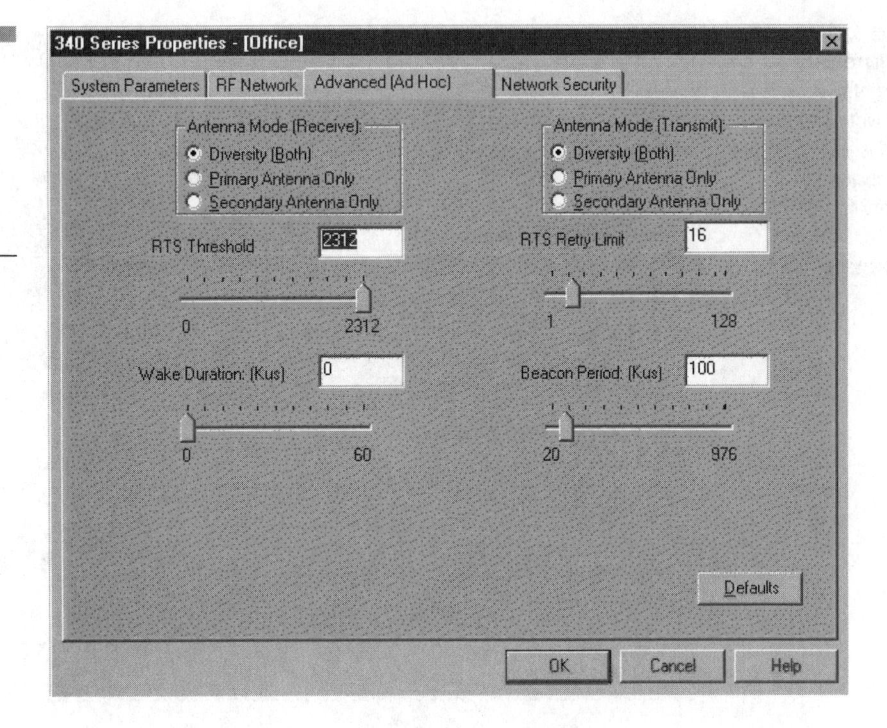

Figure 5-7 The Ad Hoc screen of Cisco's configuration utility for 340 Series client adapters

will depend on what power-saving mode was selected in the configuration utility's System Parameters screen. If the continuously awake mode was selected, the wake duration is set to 0 Kμs. If either of the two other power-saving modes were selected, the wake duration can be set from 5 to 60 Kμs. The notation Kμs is a unit of measurement indicating 1,024 microseconds, which can also be expressed as 1.024 milliseconds or .001024 seconds.

Beacon Period This setting specifies the duration between beacon packets, which are used to help clients find each other in ad hoc mode. The unit of measurement is the same as that used to set the wake duration. The range of the beacon period is 20 to 976 Kμs, with 100 Kμs set as the default.

Network Security

The Network Security screen enables the administrator to set parameters that control how the client adapter associates with an AP, authenticates to the wireless network, and encrypts and decrypts data (see Figure 5-8).

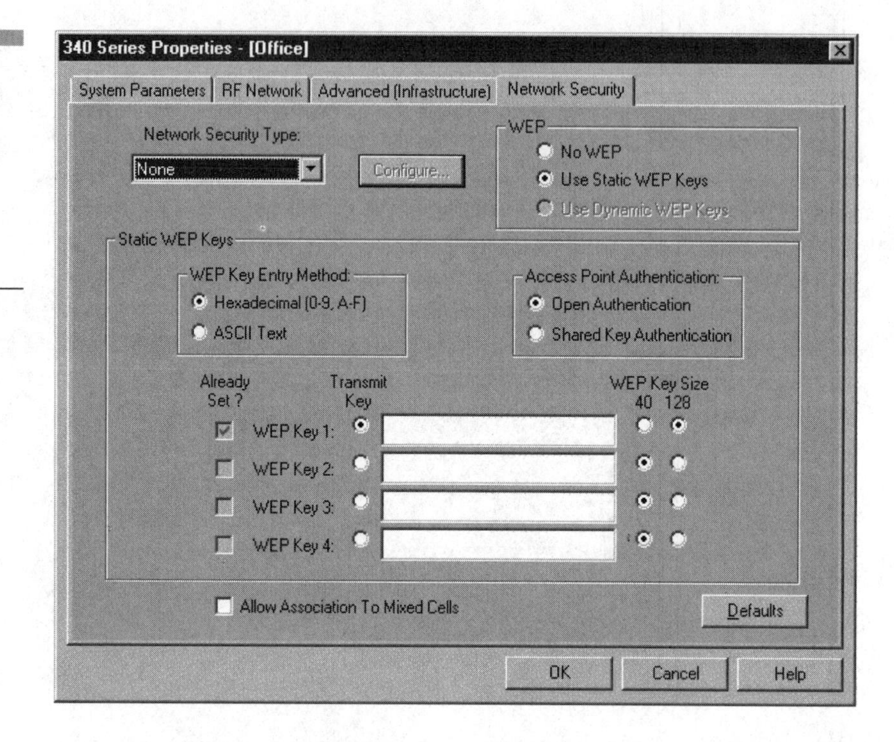

Figure 5-8 The Network Security screen of Cisco's configuration utility for 340 Series client adapters

WEP The first decision the network administrator makes is whether or not to enable WEP. As previously noted, this is not a simple decision. Because hackers have achieved success in cracking the 40/64-bit version of WEP, it can give users and administrators a false sense of security. However, this should not necessarily stop administrators from applying the 128-bit version of WEP, even though this has been cracked as well. Applying 128-bit WEP can easily stop wannabe hackers and casual snoopers.

If other choices are available that provide tighter security, they should be used instead of WEP. At this writing, a newer security standard is emerging called *802.1X*, which features the *Extensible Authentication Protocol* (EAP). It acts as the interface between a wireless client and an authentication server, such as a *Remote Authentication Dial-In User Service* (RADIUS) server, to which the AP communicates over the wired network. This is covered in more detail in Chapter 9, "Network Management."

Static WEP Keys With this setting, the WEP keys used to encrypt and decrypt transmitted data are statically associated with the client adapter and may be written in hexadecimal notation or ASCII text. If ASCII text is used, any character in the standard ASCII table can be used for the WEP key, even punctuation marks. If 40-bit WEP is selected, the keys must have 10 alphanumeric characters; if 128-bit WEP is selected, the keys must have 26 alphanumeric characters. As noted earlier, the keys must be identical for all client adapters associated with the same AP.

The client configuration tool enables multiple static WEP keys to be entered so the client device can associate with APs on different networks. Each key can be separately identified as being 40/64-bit or 128-bit WEP. If a device receives a packet that is not encrypted with the appropriate key, the device discards the packet and it is never delivered to the intended receiver.

Static WEP keys are write-only and temporary, and are lost when power to the adapter is removed or the Windows device is rebooted. Although the keys are temporary, they are not reentered with the configuration utility each time the client adapter is inserted or the Windows device is rebooted. Because the keys are encrypted and stored in the registry of the Windows device, when the driver loads and reads the client adapter's registry parameters, it also finds the static WEP keys, decrypts them, and stores them in volatile memory on the adapter.

AP Authentication This setting enables the network administrator to specify whether the client device will use open authentication or shared key authentication. As described earlier, open authentication enables the client to begin a conversation with the AP without being challenged. This mode of operation provides no security at all. Shared key authentication enables client adapters to communicate only with APs that have the same secret key.

Association to Mixed Cells This setting is used to indicate whether or not the client adapter can associate with an AP that allows both WEP and non-WEP associations. However, it is not recommended that WEP-enabled and WEP-disabled clients be allowed in the same cell because broadcast packets will be sent unencrypted, even to clients running WEP.

Conclusion

APs and clients are the key components of any WiFi network. They are available in 2.4 and 5 GHz versions. APs are available as hybrid units that accommodate both radios, enabling organizations to protect their investments in 2.4 GHz client adapters while upgrading to the higher-speed 5 GHz client adapters as the need arises.

The AP serves as the center point of a standalone wireless network or as the connection point between wireless and wired networks. In large installations, wireless users within radio range of an AP can roam throughout a facility while maintaining uninterrupted access to the network.

Both the APs and clients are available in versions for home and office environments. Office APs have more parameters to configure and are manageable via the *Simple Network Management Protocol* (SNMP). Sometimes the products permit the user to select the environment in which the product will operate. The configuration utilities of such products actually have a check box for this purpose, enabling separate profiles to be created so that a client adapter can be easily used in any environment, plus public hot spots, without having to open the configuration utility to change the settings.

CHAPTER 6

Wireless Bridges

The function of a bridge is to extend the range of a wireless link and make multiple *local area networks* (LANs) at different locations appear to users as a single network. Both functions are important, but in *Wireless Fidelity* (WiFi) networks, bridges are especially valued for their operating range. They are used by enterprises to build WiFi networks and by *wireless Internet service providers* (WISPs) to extend the reach of their services.

Bridges come in a variety of configurations and price points. At the low end, workgroup bridges extend the range of wireless connections within a building. At the midrange, bridges provide building-to-building coverage within a campus environment. At the high end, bridges extend wireless links across town or even between towns.

Although *access points* (APs) provide wireless clients with connections to wired LANs, wireless bridges connect APs to each other, expanding wireless coverage over greater distances. Bridges do not normally accept connections from clients, only from other bridges and APs (see Figure 6-1), but they can be modified to serve as APs. This has the advantage of greatly extending the range of an AP, enabling clients across town to access the wired LAN while continuing to pass traffic with other bridges.

Wireless bridges differ in their operating range. The actual range depends on many factors, including the data rate (bandwidth) desired, line

Figure 6-1 The role of a bridge in a wireless network

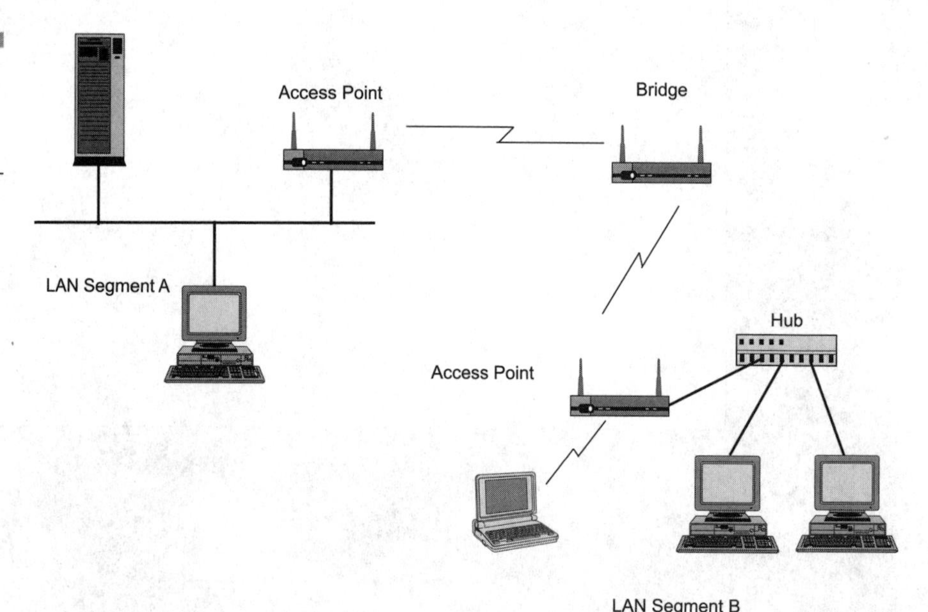

of sight, antenna type, antenna cable length, and device receiving the transmission. In an optimal installation, the range can be up to 25 miles. Ranges up to 50 miles have been reported using various combinations of off-the-shelf and custom-built components. But with greater distance comes a greater susceptibility to signal interference. Police radar and heavy machinery, for example, are intermittent sources of interference.

Another important consideration is that the bridge's antennas require not only a visual line of sight, but also a radio line of sight, which includes an elliptical region around the visual line of sight called the Fresnel zone (see Figure 6-2). For optimal performance, the Fresnel zone must be clear of all obstructions, including trees, power lines, buildings, and topographic features, such as mountains.

The bridge comes with its own *Media Access Control* (MAC) address, and the *Internet Protocol* (IP) address for a bridge is usually obtained via the *Dynamic Host ConfigurationProtocol* (DHCP). Alternatively, the network administrator can console in and set a static IP address. Bridges also support 40/64-bit and 128-bit *Wired Equivalent Privacy* (WEP), which encrypts the payload of packets sent across a radio link. The WEP key is a user-defined string of characters used to encrypt and decrypt data.

Since wireless bridges are usually placed outdoors, commercial versions usually come in moisture-proof enclosures and feature built-in lightning protection. They support *Power over Ethernet* (PoE) in which the unused pairs in the *Category 5* (CAT 5) cable carry electrical current from an Ethernet switch to power the unit. Bridges also can be configured to support the *Simple Network Management Protocol* (SNMP), making the units easy to monitor.

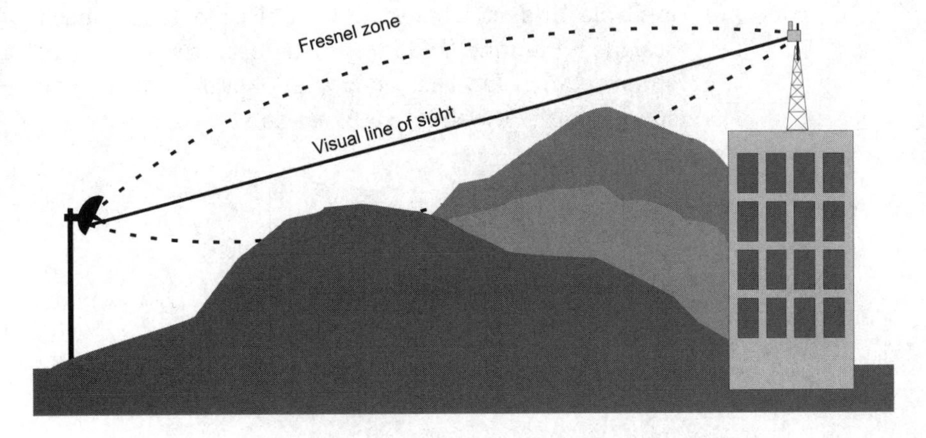

Figure 6-2 The relationship of the visual line of site to the Fresnel zone

Configuration

To configure the bridge for the first time, the console connection is used for setting up initial configuration information. This is accomplished by attaching the console port cable to the console port on the bridge. The other end of the console cable is attached to the serial port on a terminal or PC running a terminal emulation program. The terminal is usually set to 9600 baud, 8 data bits, no parity, 1 stop bit, and Xon/Xoff flow control. Future changes to the bridge can be made over the wireless link, including firmware upgrades.

Service Set Identifier (SSID)

The console port is used to set the *service set identifier* (SSID), which enables the wireless bridge to communicate with other nodes. The SSID is a unique, case-sensitive identifier of up to 32 alphanumeric characters and is attached to selected packets sent out over the radio network. It can be a plaintext description for easy identification, such as Cisco_Aironet. All devices on the same wireless network must have the same SSID or their association requests will be ignored. A unique SSID is assigned to the bridge, and if more than one bridge exists on the network, they must all use the same SSID.

IP Address

To enable remote access to the bridge using Telnet, the *Hypertext Transfer Protocol* (HTTP), or SNMP, the bridge must be assigned an IP address. If multiple bridges are on the wireless network, each of them must be assigned a unique IP address. The network administrator can also assign other detailed Internet addressing options, such as the gateway address, subnet mask, or SNMP configuration.

DHCP or BOOTP

By default, the bridge is configured to attempt to get a DHCP or *Bootstrap Protocol* (BOOTP) server to assign it an IP address. With BOOTP, IP addresses are assigned based on MAC addresses. With DHCP, IP addresses are "leased" for predetermined periods of time.

On power up, the bridge will issue boot protocol requests to see if any BOOTP or DHCP servers are on the network. If a response is received, an IP address is assigned by the server. If multiple responses are received, the bridge will pick a DHCP server over a BOOTP server. If no response is made, the request is repeated up to 30 times. If still no response occurs, the unit gives up. It is recommended, however, that a static IP address be assigned to the bridge to simplify network management and prevent delays in receiving an address through DHCP or BOOTP.

Subnet Mask

If the network uses subnets, a default gateway and an IP subnet mask will be needed for the bridge. The subnetwork mask determines the portion of the IP address that represents the subnet ID, which is used to group devices based on the network topology.

Domain Name Servers

A domain name server enables the network administrator to specify the plain language name of a known host rather than its raw IP address when accessing another node on the network. The IP addresses of the primary and backup domain name servers are entered into the appropriate fields using the bridge's configuration utility.

IP Routing Table

The IP routing table controls how IP packets originating from the bridge will be forwarded. If the destination IP address matches a host entry in the table exactly, the packet will be forwarded to the MAC address corresponding to the next-hop IP address from the table entry. If the destination is in the local subnet, the *Address Resolution Protocol* (ARP) will be used to determine the node's MAC address.

If the destination address is on another subnet and matches the infrastructure portion of a net entry in the table, using the associated subnet mask, the packet will be forwarded to the MAC address corresponding to the next-hop IP address from the table entry. If the destination address is on another subnet and does not match any entry in the table, the packet will be forwarded to the MAC address corresponding to the default gateway's IP address.

Root Configuration

The default setting for a bridge is as a root bridge. A root bridge is located at the top, or starting point, of a wireless infrastructure. The root bridge is usually connected to the main wired backbone LAN. Since the radio traffic from the other bridges' LANs will pass through this unit, the root unit is usually connected to the LAN that originates or receives the most traffic.

Before performing a detailed configuration of a bridge, the network administrator should determine whether it will be a root bridge or a nonroot bridge. A nonroot bridge, sometimes referred to as a remote or repeater bridge, establishes a connection to the root bridge or another repeater bridge to make the wired LAN to which it is connected part of the bridged LAN. Changing the default root bridge simply involves accessing the bridge's configuration utility using a terminal emulator or browser and selecting on or off for the Enable root mode setting.

After the bridge is configured as a root or nonroot unit, the network administrator can quit the terminal emulator or browser, disconnect the console port cable, and install the bridge in its desired location. After installation, an antenna alignment test should be run before proceeding with remote configuration. This test can be run from a utility installed on a client or from a *wireless LAN* (WLAN) management system. The antenna test displays signal strength and signal quality, in a text or graphical format (see Table 6-1). The test can be used to verify the link to each radio partner or to monitor signal strength while aligning directional antennas between two nodes. As the antennas are moved, the signal strength can be monitored to achieve the highest level.

The test works by sending a packet once per second to the parent AP. This packet is echoed back to the bridge, which records and displays the RF signal strength associated with that particular node. The network administrator can choose the duration of the test, from 15 to 60 seconds.

The columns provide the following information:

- **ID** This is the sequence number of the data sample. The most recent sample appears at the top of the column. In this test, 15 data samples were used.
- **Name** This is the system name of each device in the antenna alignment test.
- **Address** This is the MAC address of the device involved in the alignment test.

Table 6-1

Sample antenna alignment test results

ID	Name	Address	Signal Strength	Signal Quality
15	North Bridge	00409631158c0	100% −10 dBm	100%
14	North Bridge	00409631158c0	100% −10 dBm	100%
13	North Bridge	00409631158c0	100% −10 dBm	95%
12	North Bridge	00409631158c0	100% −10 dBm	97%
11	North Bridge	00409631158c0	100% −10 dBm	100%
10	North Bridge	00409631158c0	100% −10 dBm	99%
9	North Bridge	00409631158c0	100% −10 dBm	100%
8	North Bridge	00409631158c0	100% −10 dBm	100%
7	North Bridge	00409631158c0	100% −10 dBm	97%
6	North Bridge	00409631158c0	100% −10 dBm	100%
5	North Bridge	00409631158c0	100% −10 dBm	100%
4	North Bridge	00409631158c0	100% −10 dBm	100%
3	North Bridge	00409631158c0	100% −10 dBm	100%
2	North Bridge	00409631158c0	100% −10 dBm	100%
1	North Bridge	00409631158c0	100% −10 dBm	100%

- **Signal strength** The left side of this column displays the percentage of signal strength between the bridge and the other device; the right side of the column displays the signal strength in dBm.
- **Signal quality** This column shows the quality of the signal link between the bridge and the other device.

Data Rate

The network administrator can define the rate at which the bridge is allowed to receive and transmit data. When a repeater associates to a root bridge, data is usually transmitted between the units at the highest rate that they both support. The units can also downshift to use lower common rates if conditions warrant it. The basic rates option is set on the root bridge. All nodes in the radio cell must support this set of rates or they will not be allowed to associate.

The lowest basic rate is used to transmit all broadcast and multicast traffic as well as any association control packets. Using the lowest rate helps ensure they will be received by all nodes even at the farthest distances. The highest basic rate determines the maximum rate at which an acknowledge packet may be transmitted.

Frequency

The actual frequency allowed depends on the regulatory body that controls the radio spectrum in the location in which the unit is used. If the setting is left as auto, the unit will sample all the allowed frequencies when it is first started and try to pick one that is not in use. This setting is only allowed on the root unit since it is in charge of setting up the radio cell.

Range

Since the radio link between bridges can be quite long and vary between individual devices, the time it takes for the radio signal to travel between the radios can become significant. The range parameter is used to adjust the various timers used in the radio protocol to account for the extra delay. The parameter is only entered on the root bridge, which will inform all the other units. The range is entered as the distance in kilometers of the longest radio link in the set of bridges.

Beacon Interval

The beacon interval is expressed as Kμs, which is 1,024 microseconds, or a kilo-microsecond. The beacon packets are primarily used for radio network synchronization. A small beacon period means faster response for roaming nodes.

Broadcast SSID

When the bridge is configured as an AP, this option controls whether client nodes will be allowed to associate with the unit if they specify the empty or broadcast SSID. Clients that do not know the SSID of the bridge can transmit packets with the broadcast SSID. Any bridges present will respond with a packet showing their SSID. The client will then adopt the SSID and associate with that device. However, if the network administrator wants to ensure that clients know the SSID beforehand, this function can be disabled.

RTS/CTS

This parameter determines the minimum size of the transmitted packet that will use the *Request to Send / Clear to Send* (RTS/CTS) protocol. This value can be set from 100 to 2,048 bytes. This protocol is most useful in networks where the mobile nodes may roam far enough so the nodes on one side of the cell cannot hear the transmission of the nodes on the other side of the cell.

Before the transmission of real data, a small packet is sent out (RTS). The destination node must respond with another small packet (CTS) before the originator may send the real data packet. A node at the far end of a cell will see the RTS to or from the bridge or the CTS to or from the bridge. The node will know how long to block its transmitter to enable the real packet to be received by the bridge.

The RTS and CTS are small and, if lost in a collision, they can be retried more quickly and with less overhead than if a larger packet containing real data must be retried. The downside of using RTS/CTS is that for each data packet transmitted, another packet must be transmitted and received, which affects throughput.

Packet Encryption

This parameter controls the use of encryption on the data packet transmitted over the wireless link, which is usually 40/64-bit or 128-bit WEP, but depending on the vendor it may also include 256-bit enhanced WEP. The packets are encrypted using any one of up to four known keys. Each node in the radio cell must know all the keys in use. Not only must all nodes in the wireless network know the keys in use, but they must also be entered into the fields of the configuration utility in the same order.

Only one of these keys is used for transmitting data. This key is selected by the network administrator. Each wireless device is capable of decrypting received packets sent with any of the four keys. Any unencrypted data received by these devices will be discarded. Of course, the network administrator can also choose not to encrypt data, but this would leave transmissions exposed to hackers.

A root bridge can be set to accept associations from clients that have encryption enabled or disabled. A situation where this might be used is in the mixed vendor environment to ensure compatibility between devices. But this mixed mode of operation is not recommended for security reasons. If a client with encryption enabled sends a multicast packet to its parent, the packet will be encrypted. The parent will then decrypt the packet and retransmit it to the cell for all other nodes to see. If a hacker sees a packet in both encrypted and unencrypted form, this can make it much easier to break a key.

Time

The bridge can be configured to query a network time server so that any logs can reference the current date and time. Since the time returned by the network time server is based on *Greenwich Mean Time* (GMT), the network administrator must adjust this parameter for display in an appropriate local format. Another option is used to select whether a particular time zone uses daylight saving time so that the change is implemented automatically.

Mobile IP

A bridge can be set up to allow client roaming across different IP subnets while maintaining their original IP address. This arrangement requires that a Mobile IP stack be set up on both the bridge and the client devices. Each client is assigned an IP address and a *home agent* (HA) IP address by the network administrator. The HA resides on the subnet for which the client's IP address is local.

When the client roams to a foreign subnet, it contacts a foreign agent on that subnet, supplying its HA address. The foreign agent contacts the HA with the client's information. The HA relays any packets found on its local LAN destined to the client's IP address: first to the foreign agent and then back to the client.

From the HA, the IP addresses of mobile clients that are currently away from their home network can be displayed . For each IP address, the foreign agent it is connected with is also displayed. From the foreign agent, the IP addresses of the mobile clients that are currently visiting the agent are displayed.

Before a node is allowed to roam, the HA must be given information about the clients to validate their identity. The configuration utility asks for a range of IP addresses. The network administrator only needs to enter the low and high IP addresses in the range. If an employee leaves the company, the network administrator can remove only the IP address of the affected client device from the table. For security, the setup packets sent between the HA, foreign agent, and the clients can be encrypted. The configuration utility also enables the network administrator to set parameters that control the operation of the agents. These parameters include the following:

- **Lifetime** This parameter has two functions. It is the maximum amount of time the HA will grant a mobile client to be registered on a foreign network before renewing its registration. The lifetime value is also placed in the agent advertisement packets. Mobile clients typically use this field from the advertisements to generate the lifetime value for the registration request.
- **Replay protection** This option determines the scheme used to prevent attacks based on capturing packets and playing them back at a later time. Two replay protection methods are specified for Mobile IP: timestamps (mandatory) and nonces (optional). Nonces are a type of handshake that checks the validity of information between the mobile client's agent and a bridge's HA before granting a registration request.

  With timestamps, mobile clients use the date and time of day in the identification field of the registration request. The HA rejects the registration request if the timestamp is not close enough to its current time. A rejection message includes the HA's current time, so the mobile client can synchronize its clock accordingly.

  With nonces (random numbers), the identification field is subdivided into lower and higher halves. The registration request specifies to the HA which value to place in the lower half of the registration reply. The registration reply specifies to the *mobility agent* (MA) which value to place in the upper half of the next registration request. Both sides check nonces; if a nonmatching registration message is received, the MA ignores the message and the HA rejects the message. It also sends back a message that includes values it expects in the next registration request. This process is summarized in Figure 6-3.
- **Broadcasts** Mobile clients can be configured so that broadcasts from their home network will be forwarded to them via tunneling. Some protocols, such as the *Network Basic Input/Output System* (NetBIOS), require broadcast packets from the home network to maintain proper operation. Unless needed, however, this option should be disabled to avoid unnecessary traffic over the wireless links.

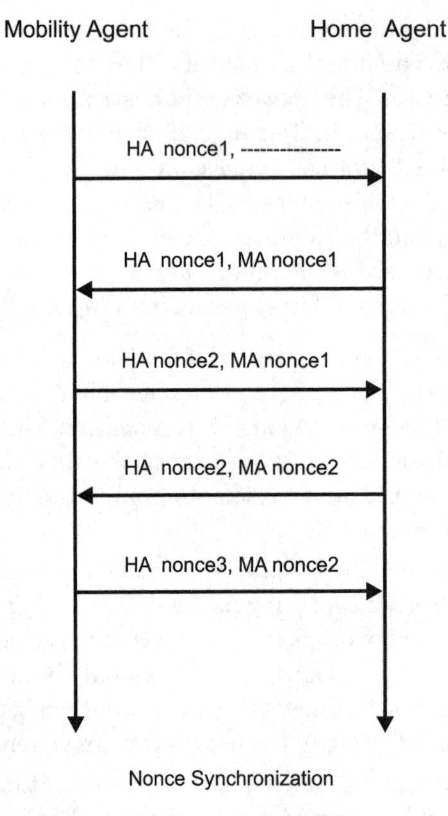

Figure 6-3 Replay protection via the nonces handshake between an MA and an HA

- **Registration required** Mobile clients can be set up to enable registration to an HA without the use of a foreign agent via a *care of address* (COA) dynamically acquired while on the foreign network. This is useful in cases where foreign agents have not yet been deployed on the foreign network, but this scheme consumes IP addresses on that network. Unless plenty of spare IP addresses exist, mobile clients should be forced to always register using a foreign agent.
- **Host redirects** This indicates whether or not the foreign agent can send an *Internet Control Message Protocol* (ICMP) packet to mobile clients registered through it, specifying the IP address of a router for the mobile client to use. Disabling this feature will result in the mobile client always using the foreign agent as its default gateway (router). Enabling this feature may improve performance while visiting a foreign network, but it may also pose connectivity problems caused by ARP broadcasts from the mobile node.

- **Control agent advertisements** Agents advertise themselves on the LAN so that mobile clients can find them and determine whether they are home or away. The network administrator can specify how frequently (in seconds) the MA will send out an ICMP router advertisement multicast. These advertisements are used by the mobile clients to locate the MAs and determine to which network they are currently attached. The more frequent the advertisement, the faster the mobile node will become aware that it has attached to a new network and can start the registration or deregistration process.

Spanning Tree Protocol (STP)

In multibridge environments, the *Spanning Tree Protocol* (STP) is used to remove loops so that packets do not circulate endlessly without ever reaching their destination. STP is a standard feature in conventional bridges used on cabled LANs, but it is just as applicable to bridges in WLANs. In essence, STP enables the bridges in an arbitrarily connected infrastructure to discover a topology that is loop free (a tree) and ensures that a path exists between every pair of LANs (a spanning tree).

If more than one path exists from one LAN to another, the infrastructure contains a loop. If Node A transmits a multicast packet, both Bridge 1 and Bridge 2 will try to forward the packet to LAN 2. Each bridge, on seeing the other's transmission on LAN 2, will forward the packet back to LAN 1. The cycle will continue and the packet will loop forever, taking up all the bandwidth of the bridges (see Figure 6-4).

STP works by having the bridges transmit special configuration messages to each other.[1] The messages contain information that enables the bridges to

- Elect a single bridge from all the bridges on all the LANs to be the root bridge. Each of the other bridges then calculates the distance of the shortest path to the root bridge.
- Find a bridge for each LAN that is closest to the root bridge.
- Select its own port to be the root port. This bridge has the best path to the root bridge.

[1]STP runs only on the root bridge, not on repeaters. Repeaters only transmit packets or change state on commands from the root bridge.

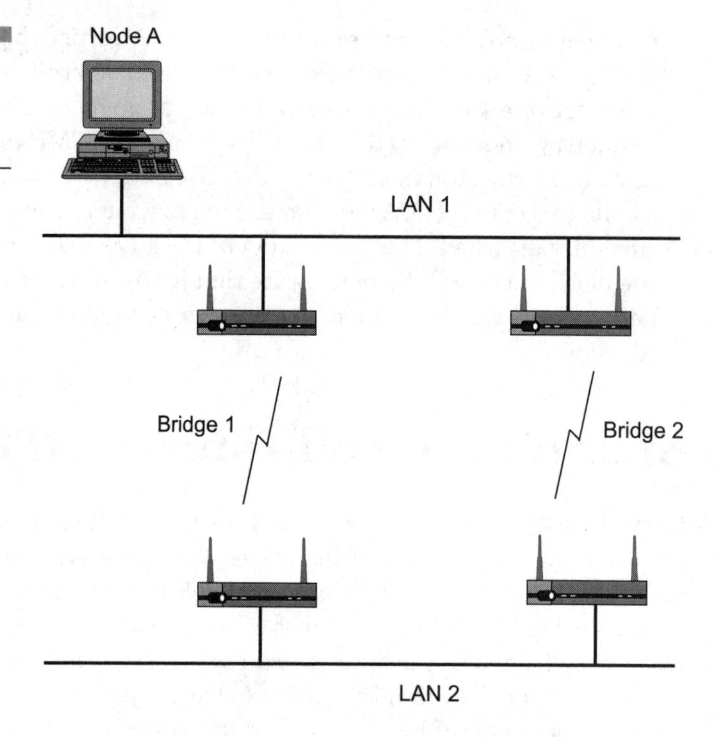

Figure 6-4 An example of a loop on a bridge infrastructure

- Select ports that are included in the spanning tree. Ports are included if they are root ports or the bridge itself has been selected as the designated bridge for the LAN.

All ports on a bridge, either the root port or the designated port for their LAN, are allowed to forward packets. All others are blocked and do not transmit or receive any data packets.

It may take some time for the protocol to determine a stable loop-free topology because of the time it takes for messages to pass from one end of the infrastructure to the other. If the ports are allowed to forward packets while the protocol is stabilizing, then temporary loops could form. To avoid temporary loops, ports are not allowed to go immediately from the blocked state to the forwarding state. Instead, the ports must first go through a "listening" state whereby they may receive and transmit configuration messages, but must block all data traffic. The amount of time spent in the listening state must be at least twice the end-to-end transmit time of the infrastructure.

If a port is still part of the spanning tree at the end of the listening period, it is put in the "learning" state whereby it can still receive and transmit configuration messages, but it is also allowed to learn the source addresses from the packets received from its LAN. At this point, however, it is still not allowed to forward any packets. The learning state is used to decrease the amount of flooding of unknown destination addresses that would occur if the port started forwarding before any entries were made in its learning table. Once the learning period is over, the port is allowed to forward data normally.

Several parameters must be set in the bridge's configuration utility to get STP working properly. With a small infrastructure consisting of a point-to-point wireless link between two units, loops will never take place. In such cases, the STP can be set to off. If the network administrator is not really sure how many bridges are out there, no harm exists in changing the setting to on because the amount of overhead traffic between bridges is small.

Priority

This option is used to set the priority value appended to the infrastructure address of the bridge ID. By changing the priority value, the network administrator can influence which bridge in the infrastructure will become the root bridge. The lower the priority value, the more likely the bridge will be the root. If all other bridges are set to the default value, a bridge set with a lower value will become the root.

Hello Time

This option is used to set the interval time, in seconds, between transmitted configuration messages. This value is only used if the local bridge becomes the spanning tree root bridge. If the interval time is set too high, the infrastructure will respond slowly in resolving any conflict problems; if it is set too low, the infrastructure will be congested with hello message traffic. The interval time values range between 1 and 10 seconds.

Forwarding Delay

This option is used to set the delay time, in seconds, that the ports will spend in the listening and learning states. This value is only used if the local bridge becomes the spanning tree root bridge.

This option is also the timeout period used to age learned addresses whenever the spanning tree topology changes. The value should be at least twice the transit time of a packet sent from one end of the infrastructure to the other. This allows news of a topology change to reach all nodes, and it allows all ports to be blocked before new ports enter the forwarding state.

If the interval time is set too low, then temporary loops could be formed; if it is set too high, it will take longer for the infrastructure to become active after a spanning tree topology change has been made. The delay time values range between 4 and 30 seconds.

Message Timeout

This option is used to set the timeout period, in seconds, that a blocked or root port watches for configuration messages from the infrastructure's designated port. This value is only used if the local bridge becomes the spanning tree bridge. Each time a configuration message is received, the timer is started. If the timer expires, the root bridge is assumed to have failed and the spanning tree infrastructure will be reconfigured.

If the timeout period is set too low, the spanning tree infrastructure may reconfigure itself unnecessarily and messages can be lost due to heavy traffic on the infrastructure. If it is set too high, the infrastructure will take longer than necessary to recover from failed ports or bridges. The upper limit on the allowed range is determined by the setting of the forwarding delay. The timeout period must be less than twice the forwarding delay, minus 1 second. Therefore, the timeout values range between 6 and 29 seconds.

Port

This option is used to enable STP on the local port. The default setting is on, which enables all root bridge LAN ports to be initially placed in the listening state. If the option is turned off, the LAN ports are placed in the forwarding state. If the port's LAN will always be connected to the bridge and loops will never occur, turning the protocol off will prevent the port from transmitting configuration messages on every timeout period.

Local Port Priority

This option is only used when two or more repeaters are connected to the same LAN for redundancy and the network administrator wants to select which one will forward the packets. The port assigned the lowest priority value will be the one to forward.

Local Port Cost

The value for the cost option is added to the root cost field from any received configuration messages to determine if the port has the least cost path value to the root. The larger the cost value, the more likely the port will be a backup for another active port on its LAN. If no port is active, it is likely the LAN will be a leaf of the infrastructure tree or a less used LAN in the tree.

Ports of Active Connected Repeaters

The port, priority, and cost options are used to configure the ports of active and connected repeaters in the root's radio tree. These parameters are the same as those described earlier, except that when the values are entered, the network administrator will be prompted for the applicable port number, which can be obtained from the port ID field on the configuration tool's protocol status display screen.

Protocol Status

This display shows the overall status of the STP and the state of each port on the local bridge. The following information that be displayed from the status screen:

- *Bridge ID* shows the ID of the local bridge.
- *Root ID* shows the ID of the spanning tree root. If the local bridge is not the root, then the cost to the root is also displayed.
- *Topology change* shows whether the short aging timeout is currently in use because of a port state change somewhere on the infrastructure.

- *Network hello interval, network forward delay, and network message timeout* show the timeout values received from the root bridge in use by all bridges on the infrastructure. These values override any locally configured values.
- *Port address* shows the infrastructure address of the bridge on which the port resides.
- *ID* shows the port ID, which consists of the port priority and the port number. As each repeater connects to the root, its port is assigned the next available port number.
- *Cost* shows the cost for the port as configured by the network administrator.
- *State* shows the current state of the port if used: forward, learn, listen, or blocked.
- *Type* shows the current port type as root, designated, or blocked.
- *Designated (bridge, port, root cost)* shows the designated bridge and port for the specific LAN, as well as the cost to the root from the designated port.

Performance Monitoring

The monitoring tools of bridges display a number of statistics about their performance that network administrators can use to fine-tune the network or head off problems. Such tools can be used to gather baseline statistics when the network is considered to be operating at peak performance. These baseline statistics are saved for comparison against new measurements taken periodically. If significant negative deviation takes place, steps can be taken to determine the cause and find an appropriate remedy.

Throughput

This display shows details about the data packets passing through a particular bridge. The statistics are categorized in terms of radio transmit and receive, and bridge transmit and receive, providing the following information:

- *Recent rate* displays the event rates, per second, averaged over the last 10 seconds.
- *Total* displays the number of events that have occurred since the statistics were last cleared.

- *Average rate* displays the average event rates, per second, since the statistics were last cleared.
- *Highest rate* displays the highest rate recorded since the statistics were last cleared.
- *Packets* displays the number of packets transmitted or received.
- *Bytes* displays the total number of data bytes in all the packets transmitted or received.
- *Filtered* displays the number of packets that were discarded as a result of an address filter being set up.
- *Errors* displays the number of errors that may have occurred.

Radio Errors

This display shows the extent of radio receiver and transmitter errors that have occurred at the bridge:

- *Buffer full frames lost* shows the number of frames lost due to a lack of buffer space in the unit.
- *Duplicate frames* shows the number of frames received more than once. This is usually due to a frame acknowledgment being lost.
- *CRC errors* shows the number of frames received with an invalid *Cyclic Redundancy Check* (CRC), which is usually caused by interference from nearby radio traffic. CRC errors can also occur from random noise when the receiver is idle.
- *Retries* shows a cumulative count of the number of times a frame had to be retransmitted due to an acknowledgment not being received.
- *Maximum retries/frame* shows the maximum number of times any one frame had to be retransmitted. An excessive number of retries may indicate a poor-quality radio link.
- *Queue full discards* shows the number of times a packet was not transmitted due to too many retries occurring to the same destination. This occurs if packets destined to the same are taking up more than their share of transmit buffers.

Ethernet Error Statistics

This display provides details on the receiver and transmitter errors that have occurred at the bridge:

- *Buffer full frames lost* shows the number of frames lost due to a lack of receiver buffer space in the unit.
- *CRC errors* shows the number of frames received with an invalid CRC.
- *Collisions* shows the number of times a collision occurred while the frame was being received. This would indicate a hardware problem with an Ethernet node somewhere on the infrastructure.
- *Frame alignment errors* shows the number of frames received whose size in bits was not a multiple of 8. Occasionally, extra bits of data are inadvertently attached to a transmitted packet causing a frame alignment error.
- *Over-length frames* shows the number of frames received that are longer than the configured maximum packet size.
- *Short frames* shows the number of frames received that are shorter than the allowed minimum packet size of 64 bytes.
- *Overruns* shows the number of times the hardware receives *first in, first out* (FIFO) overflow. This should be a rare occurrence.
- *Misses* shows the number of Ethernet packets that were lost due to lack of buffer space on the unit.
- *Excessive collisions* shows the number of times transmissions failed due to excessive collisions. This usually indicates the frame had to be continuously retried due to heavy traffic on the Ethernet infrastructure.
- *Deferrals* shows the number of times frames had to wait before transmitting due to heavy traffic on the cable.
- *Excessive deferrals* shows the number of times the frame failed to transmit due to excessive deferrals. This usually indicates the frame had to be continuously retried due to heavy traffic on the Ethernet infrastructure.
- *No carrier sense present* shows the number of times the carrier was not present when a transmission was started. This usually indicates a problem with a cable on the Ethernet infrastructure.

- *Carrier sense lost* shows the number of times the carrier was lost during a transmission. This usually indicates a problem with a cable on the Ethernet infrastructure.
- *Out of window collisions* shows the number of times a collision occurred after the 64th byte of a frame was transmitted. This usually indicates a problem with a cable on the Ethernet infrastructure.
- *Underruns* shows the number of times the hardware transmit FIFO buffer became empty during transmission. This should be a rare occurrence.
- *Bad length* shows the number of times an attempt was made to transmit a packet larger than the specified maximum allowed.

Applying Filters

If the bridge is connected to an infrastructure with a large amount of multiprotocol traffic, the network administrator may be able to reduce the amount of radio traffic by blocking out (filtering) those addresses or protocols that are not needed. Such filtering is especially important for battery-operated radio nodes, which might otherwise have to waste battery power receiving irrelevant multicast messages that will only be discarded. To achieve consistent performance throughout the infrastructure, especially as clients roam, any filters set for one bridge should be duplicated on all bridges.

The network administrator can control the direction packets are traveling before they are affected by the filters. For example, only packets from the LAN might have filters applied, while packets from the radio will not be filtered. This option reduces the amount of LAN traffic to the radio network. Alternatively, packets in both directions can be filtered. This option enables control of the type of traffic the radio nodes may use.

Filtering Multicast Addresses

Using the bridge's configuration utility, the network administrator can control the filtering of multicasts based on whether or not multicast addresses appear as entries in the bridge's table:

- **Discard** This option prevents multicasts with no table entries from being forwarded out to the radio network.
- **Forward** This option enables multicasts with no table entries to be forwarded out to the radio network.
- **Access point** This option enables multicasts with no table entry to be forwarded only to other APs and bridges, but not to the clients.
- **Non-power-saving protocol** This option enables multicasts with no table entries to be forwarded out to the radio network to non-power-saving end nodes, but not to any nodes using the *Power-Saving Protocol* (PSP).

If the network administrator wants to filter special multicast addresses differently than the defaults described previously, the option is available to add or remove a multicast filter. If the network administrator knows that the radio nodes are not going to communicate with each other but will only communicate with nodes on the wired LAN, multicasts received from the radio nodes can be set so they are not rebroadcast to the radio cell but are forwarded only to the wired LAN. For example, if a large number of radio clients only talk to the network server, enabling multicast filtering will result in much less radio traffic congestion.

Filtering by Node Addresses

The forwarding of packets can be controlled based on the source node addresses. As with multicast filtering, a default action exists for those addresses not in the table. The network administrator can enter actions for specific addresses to override the default action. Specific node filters can be entered as either the 6-byte infrastructure address of the node or by its IP address. If the latter method is used, the bridge will determine the infrastructure address associated with the IP address and use this for the actual filtering.

Packets may be filtered based on the source address in the received packet. For example, if the network administrator wants to prevent all but a limited number of clients to communicate with nodes on the radio network, the default action would be set to discard and then entries would be added for the specific clients whose action is forward.

The bridge is always performing filtering based on the destination MAC address of the packets it receives. The bridge will learn where a node is based on the source address of received packets and then make a decision as to whether to forward a packet based on its knowledge of the location of the node.

Default actions can be set when doing destination address filtering. The Ethernet-destination value specifies the default action for packets received on the Ethernet. The radio-destination action specifies the default action for a packet received on the radio interface. The allowed values for each are discard or forward.

Bridges usually come with source address filtering turned off by default. This saves processing power since the unit has to look up the source address of each incoming packet to see if a filter is to be applied. Once a decision is made by the network administrator on which filters to apply, individual source filters can be made active with the forward or discard setting.

The network administrator can add filters for specific addresses to the filter table by entering either the infrastructure address or IP address of the node to which the filter applies. The network administrator specifies whether this is a source address, radio destination address, or Ethernet destination address filter. Then the filter action is applied to this address: forward, discard, or remove the filter. When a node address filter is entered by the IP address, the bridge first determines the infrastructure address associated with this IP address. The actual filtering is done based on the infrastructure address. One or all specific node filters can be removed by specifying a single node's infrastructure address or a single node's IP address, or by entering an all parameter.

Filtering by Protocol

Traffic can be filtered based on the type of protocol used to encapsulate the data in the packet. This type of filtering can have the most value in almost all situations and is the preferred method of filtering. With this type of filtering, the bridge can be set to forward only those protocols that are being used by the remote radio nodes. This is easier than setting up filters based on infrastructure and IP addresses.

The bridge can be set up to monitor and record the list of protocols currently being forwarded, how many packets were encountered, and whether the packet came from the LAN or the radio. To set up the protocol filters, the network administrator starts the monitor and lets it run. Filters are added by selecting the protocols from the monitor list.

A default action can be assigned for those protocols not in the list of explicitly filtered protocols. If the network administrator knows exactly which protocols are used by the radio nodes, the default action can be set to discard, and filters are added to forward only those protocols that will be used. If all the protocols that will be used by the radio nodes are not known, but the network administrator knows certain protocols will not be used, the default action can be set to forward, and filters are added to discard only those protocols that will not be used.

Once a filter has been added for the IP protocol, the network administrator can also filter packets based on their *User Datagram Protocol* (UDP) or *Transmission Control Protocol* (TCP) port number, their IP subprotocol (UDP, TCP, or ICMP), or an IP address range.

Logs

The bridge produces logs that record the occurrence of significant events occurring within the unit and on the infrastructure. The following types of logs are available:

- **Information logs** This type of log records status changes that occur in the normal operation of the system, such as when an end node associates to a bridge.
- **Error logs** This type of log records errors that occur occasionally, but that the unit can easily recover from, such as errors that occur during the reception and transmission of packets to and from the unit.
- **Severe error logs** This type of log records errors that drastically affect the operation of the system. Although the system will continue to run, action is required by the network administrator to return the unit to normal operation.

All logs are stored within the bridge's memory. Log entries are displayed in least recent to most recent order. If the memory buffer becomes full, the oldest log in the buffer will be replaced by the most recent. Only logs that have occurred since the unit was last powered up or since the memory buffer was cleared will be saved.

Setting SNMP Traps

The SNMP standard provides a limited number of unsolicited messages called *traps*, which are sent from a device to an SNMP application. These messages can be sent by the SNMP agent on the device to notify an SNMP application of a change in the status of the device. Bridges occupy a key position in the wireless infrastructure and may be set up to generate SNMP traps and send them to a network management station. A trap is sent whenever a significant event occurs. The following trap messages will be sent as they occur:

- A cold-start trap will be sent when the bridge is first powered up.
- A link-up trap is sent for a bridge as soon as the radio is configured.
- A link-down trap is sent when a bridge configuration changes or encounters a severe error condition.
- A link-up trap is sent when a bridge's configuration is changed or restored from a severe error condition.
- An authentication failure trap will be sent if an SNMP request is received with an unknown community name.

In addition, enterprise-specific traps can be sent whenever a log of a given severity or higher is produced. The generated trap will contain the text of the log message along with the severity of the log.

The network administrator can choose one bridge to monitor and have all other units associate with that unit as their host for logs. All logs can be forwarded to a UNIX host so that if the bridge fails for any reason, the logs may still be viewed on the UNIX host. At the UNIX host, the current time and IP address of the unit that sent the log will be added.

Conclusion

Wireless bridges are used to connect two or more networks, typically located in different buildings, providing high data rates for data-intensive applications. They connect discrete sites into a single LAN, even when they are separated by obstacles such as freeways, railroads, and bodies of water.

Remote bridges connect hard-to-wire sites, noncontiguous floors, satellite offices, corporate campus settings, temporary networks, and warehouses. Wireless bridges also enable multiple sites to share a single, high-speed connection to the Internet. The high-speed links between wireless bridges deliver throughput faster than T1/E1 lines for a fraction of the cost, eliminating the need for expensive leased lines or fiber-optic cable. The initial hardware investment can be quickly paid for with the money saved on leased-line service. The systems are easy to install and configure, they are compact and unobtrusive, and they can be redeployed quickly as network requirements or company locations change—without any involvement from the local telephone company. They also do not require an FCC license, even when the signals travel distances of 25 miles or more.

CHAPTER 7

Wireless Security

Organizations of every type and size are considering, if not deploying, *wireless local area networks* (WLANs). The demand for wireless access to LANs is fueled by the growth of mobile computing devices, such as laptops and *personal digital assistants* (PDAs), and users' desire for easy access to the network without having to set up a cable connection. With WLANs becoming universally accepted, *information technology* (IT) managers are demanding effective security that encompasses access control and privacy. Access control ensures that only authorized users can enter the corporate network. Privacy ensures that transmitted data can be received and understood only by the intended audience.

Although the *Institute of Electrical and Electronics Engineers* (IEEE) 802.11b standard for *Wireless Fidelity* (WiFi) networks includes components for ensuring access control and privacy, these components must be deployed on every device in a WLAN. An organization with hundreds or thousands of WLAN users needs a solid security solution that can be managed effectively from a central point of control. The lack of centralized security is the primary reason why WLAN deployments have been limited to relatively small workgroups and specialized applications.

WLAN Security

As noted in previous chapters, the IEEE 802.11b standard defines several mechanisms for providing access control and privacy on WLANs: *service set identifiers* (SSIDs), *Media Access Control* (MAC) addresses, and encryption via *Wired Equivalent Privacy* (WEP). Other methods may be used as well, such as server-based *Remote Access Dial-In User Service* (RADIUS) and a *virtual private network* (VPN), which provides encryption and runs transparently over the WLAN.

Service Set Identifiers (SSIDs)

SSIDs were designed to enable wireless networks to be better managed by ensuring that wireless clients talk only to the right *access point* (AP). By assigning the same SSID to both the AP and client devices, the scheme allows multiple *radio frequency* (RF) networks to operate in the same physical area without the risk of clients accessing the wrong network. If the SSIDs on the AP and client devices do not match, entry into the network is denied.

Although not intentionally designed to provide security, SSIDs do provide a first line of defense against attack. If the SSID is known only to those authorized to use the wireless connection, in essence it becomes a first-level security mechanism. Although it has been amply demonstrated that SSIDs can easily be learned by hackers, they should be implemented anyway, if only to prevent casual users from accessing the network. Exposing the SSID to eavesdroppers can be minimized by removing it from the AP broadcast beacon, but this method alone should never be relied upon for fail-safe security. When a client probes an AP for its SSID, the AP responds with a one-time broadcast containing the SSID. Ultimately, however, a patient attacker will still discover the SSID.

MAC Addresses

Another authentication method is based on the physical address, or MAC address, of a client. An AP will enable an association by a client only if that client's MAC address matches an address in the AP's authentication table. But relying on MAC filters is considered to be weak security because on many wireless cards it is possible to change the MAC address by using the configuration utility that comes with the wireless card. An attacker could sniff a valid MAC address from the wireless network traffic and then reconfigure his or her own card to gain access.

Another problem is that once a hacker learns how a WLAN controls access, it is possible to gain entry to the network without being noticed by stealing a valid client's access. Stealing a client's access entails the hacker mimicking the valid MAC address and using its assigned *Internet Protocol* (IP) address. The hacker waits until the legitimate system stops using the network and then takes over its identity to get back on the network. This provides the hacker with direct access to all devices on the network, even gaining access to the wider Internet, all the while appearing to be a valid user.

Yet another problem with this method of authentication is that it gets more difficult to manage as the number of clients increases. It also makes it more difficult for visiting mobile users to access the network, since the MAC address of their devices would have to be entered into the AP's authentication table. As employees leave the organization, the risk of security being compromised increases because they may have written down or remember the MAC address of their computer and could use it to regain access to the corporate network.

Encryption via WEP

Encryption is the process of passing a cleartext message through a mathematical algorithm designed to produce what is known as *ciphertext*, which makes the message unintelligible to all but the intended recipient upon decryption.

Two categories of encryption are in common use today: symmetric encryption, also known as shared-key encryption, and asymmetric encryption, also known as public/private encryption. The former is about a thousand times faster than the latter, making it useful for the bulk encryption of data. Typically, a well-designed encryption algorithm using a longer key results in a higher degree of security because more processing power is needed to try every possible key to decrypt a message. But the latest hacking tools do not necessarily need to test every key to open a message; instead, statistical sampling techniques may be used to decrypt a message well before all keys are tested.

The IEEE has specified the use of WEP as the means to encrypt 802.11 data frames. For encryption, WEP uses the RC4 stream cipher developed at RSA Security, Inc. Unlike block ciphers, which process a fixed number of bytes in each encrypt/decrypt function, the stream cipher applies the encrypt/decrypt function on a unit of plaintext: the 802.11b frame.

WEP is a symmetric encryption mechanism; that is, the key is the one piece of information that must be shared by both the encrypting and decrypting endpoints. RC4 enables the key length to be variable, up to 256 bits, as opposed to requiring the key to be fixed at a certain length. The IEEE specifies that 802.11 devices must support 40-bit keys, with the option to use longer key lengths. Many vendors now support 128-bit WEP encryption with their WiFi solutions, and a few have begun to offer 256-bit encryption.

Since WEP is a stream cipher, a mechanism must be in place to make sure that the same plaintext will not generate the same ciphertext. The IEEE allows for the use of an *initialization vector* (IV), which is concatenated with the symmetric key before generating the ciphertext. The IV is a 24-bit value that ranges from 0 to 16777215. The IEEE recommends that the IV change with every frame, but this is not a requirement. The IV must be sent to the receiver unencrypted in the header of the 802.11 data frame. The receiver can then concatenate the received IV with the WEP key it has stored locally to decrypt the data frame.

As shown in Figure 7-1, the plaintext itself is not run through the RC4 cipher; instead, the RC4 cipher is used to generate a unique keystream for the frame using the IV and WEP base key. The unique keystream that

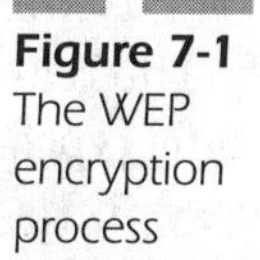

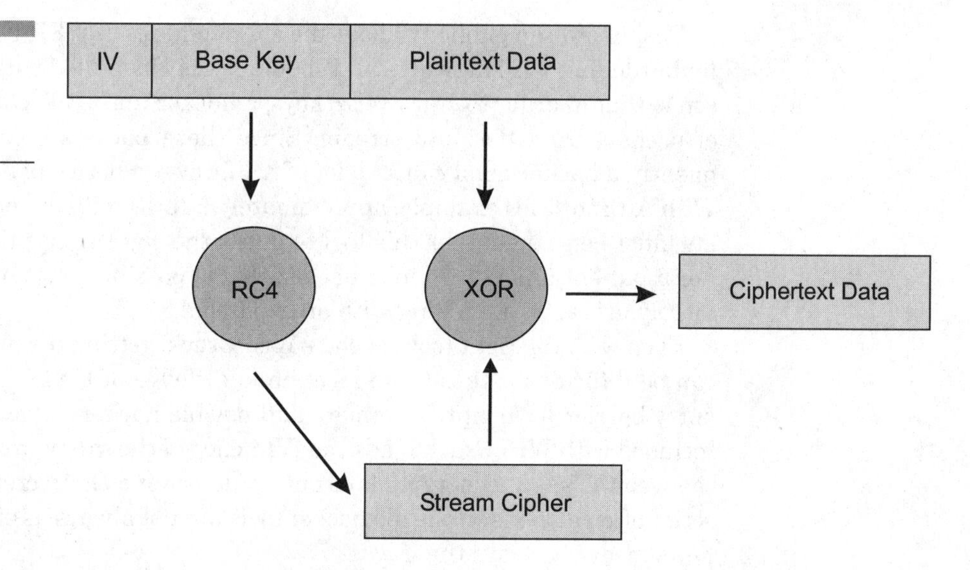

Figure 7-1 The WEP encryption process

results from this process is combined with the plaintext and run through a mathematical function called XOR, which actually produces the ciphertext.

Nothing is wrong with the RC4 algorithm used for encryption; RC4 is the popular algorithm protecting the millions of users who access secure web pages and send data via the *Secure Sockets Layer* (SSL) protocol. This protocol is secure and RC4 in SSL has never been broken. However, the way WEP actually handles the RC4 keys to be used for encrypting the 802.11 payloads compromises security. WEP produces RC4 keys that are too similar and easy to attack. WEP in its original form is flawed because it produces weak RC4 keys.

Weak RC4 keys are produced in the following way. The IV is concatenated to an RC4 key to make up the actual key that WEP uses for converting cleartext to cyphertext. With WEP, the IV is transmitted in the 802.11 payload in the cleartext, along with the cyphertext for rapid decryption at the receiving end. The IV is a sequential number that is repeated more or less frequently, depending on the amount of traffic. The repeated IV enables hackers to compare different encrypted payloads for which part of the key is known. With enough sample data, the full RC4 key can be derived, all without having to test every possible 128-bit key. In fact, tools are now readily available on the Internet that enable even novice users to successfully crack WEP keys without having significant skills or equipment. In sum, the primary design flaws that make WEP vulnerable to hackers were not addressed by the increase in key size from 40/64 bits to 128 bits.

Despite all the publicity about the shortcomings of WEP, it is still useful for hardening WiFi networks and should not be dismissed entirely. The reason is that in order to crack WEP keys, a hacker must collect specific types of packets from the data stream. Since these packets occur very infrequently, a hacker must collect a lot of traffic over a period of days or weeks. With AirSnort, for example, approximately 5 to 10 million encrypted packets must be gathered for this tool to guess the right encryption password. For a patient and determined hacker, this is possible, but not for one who merely drives by a building with an exposed AP.

Even with the right tools, such as WEPCrack, getting the program to run can be difficult for those who have no knowledge of UNIX, presenting an entry barrier for nonprogrammers and newbie hackers. One of the scripts included with WEPCrack reads the IVFile.log of the target system and uses the weak IVs, plus encrypted output, to determine the secret key, but the script often guesses wrong. So hacker tools are not always as effective as the news media portrays them.

Vendors have also improved WEP. Although many of these improvements are proprietary, they also contribute to security. RSA Security has even come up with a WEP fix that entails encrypting every packet with a distinct key, rather than reusing the same key repeatedly. Although this capability represents an important development in security over the original WEP keys, it should be noted that it is not part of any 802.11 specification at this writing, and its use would require that all users on the network use the same RSA software to realize the benefit of enhanced security.

Although the version of WEP that is used in most WiFi systems is indeed flawed, the level of knowledge and effort required to exploit these flaws is considerable, deterring all but the most motivated hackers. Even so, WEP alone should not be relied upon for total security. Unscrupulous competitive-information brokers can easily hire the required expertise and can afford the time it takes a hacker to break into a WLAN.

Authentication Mechanisms

The IEEE specifies two authentication algorithms for 802.11-based networks. First, open authentication is a null authentication algorithm because any station requesting authentication is granted access. The second form of authentication is called shared-key authentication, which requires that both the requesting and granting stations be configured with matching WEP keys. The requesting stations send an authentication request to the granting station. The granting station sends a plaintext challenge frame to the requesting station. The requesting station WEP encrypts the challenge frame and sends it back to the granting station. The granting station attempts to decrypt the frame, and if the resulting plaintext matches what the granting station originally sent, then the requesting station has a valid key and is granted access.

The flaw in this scheme is that the challenge packet is sent in the clear to the requesting station and the requesting station replies with the encrypted challenge packet. Consequently, a hacker can derive the stream cipher by analyzing both the plaintext and ciphertext. This information can be used to build a decryption dictionary for that particular WEP key.

The IEEE 802.11b standard describes WEP as an optional encryption mechanism for securing wireless data. Although the use of 40/64-bit WEP is optional, vendors are required to support it to qualify for certification by the WiFi Alliance (formerly the *Wireless Ethernet Compatibility Alliance* [WECA]). But hackers can now easily break the original version of WEP, so it does not offer effective security for the enterprise. Vendors employ other versions of WEP that use 128-bit encryption, which is not as easily broken. As noted, however, even 128-bit WEP can be broken with the hacker tool WEPCrack.

Regardless of which version of WEP is used, the 802.11b standard provides two ways of defining the WEP keys to be used on a WLAN. In the first method, a set of up to four default keys is shared by all stations—clients and APs—in a wireless subsystem. When a client obtains the default keys, that client can communicate securely with all other stations in the subsystem. The problem with default keys is that when they become widely distributed throughout the organization, they are more likely to be compromised.

With the second method of defining WEP keys, each client establishes a key-mapping relationship with another station when they are configured to operate in ad hoc mode. This is a more secure form of operation because fewer stations have the keys, but the process of distributing and managing such keys becomes more difficult as the number of stations increases.

It was originally believed that 128-bit encryption would be virtually impossible to break due to the large number of possible encryption keys. But hackers have since developed methods to break 128-bit WEP without having to try each key combination, proving that this system is not totally secure. These methods are based upon the ability to gather enough packets off the network by eavesdropping to determine the encryption key. Although WEP can be broken, it does take considerable effort and expertise to do so. It is therefore recommended that WEP be enabled and that the keys be rotated on a frequent basis. This will be made easier with the

IEEE's emerging security standard, which provides mutual authentication of a client and server through the AP.

The 802.11i standard will include *WiFi Protected Access* (WPA), which among other things generates a new key periodically during the transmission. While WEP uses fixed keys that are easy to attain via commonly available software such as Netstumbler, WPA uses the *Temporal Key Integrity Protocol* (TKIP) for generating a new key for every 10KB of data transmitted over the network.

The way WPA will work in the enterprise environment is similar to the setup of any authentication system. The clients and access points must have WPA enabled for encryption to and from an authentication server, such as a RADIUS server, with centralized access management. Once the server authenticates the user, the access point will let that user onto the wired network. Up to that point, the client had only talked to the server.

Eventually, the *Advanced Encryption Algorithm* (AES) will be incorporated into the IEEE security standard to make the encryption key unbreakable. The reason for its delay is that it requires a hardware upgrade and most users will not be able to take advantage of it with existing equipment. AES replaces the *Digital Encryption Standard* (DES), which has been in use since 1977, but in recent years has also been broken.

DES-based encryption software uses an algorithm that encodes 64-bit blocks of data and uses a 56-bit key; the length of the key imposes a difficult decoding barrier to would-be intruders because 72 quadrillion (72,000,000,000,000,000) keys are possible. The DES offers four different encryption modes. Direct mode is the easiest to implement but provides the least security because it enables independent coding of each block of a message. Independently coded blocks can develop coding patterns in lengthy transmissions; such patterns can make the encryption technique vulnerable to unauthorized access. In the other three modes, the coding of each data block varies depending on the coding of one or more previous blocks, reducing the risk of revealing a pattern in encoding that could provide clues to the decryption key.

Despite its 20-year performance of foiling decodes, DES was finally broken in 1998 by the Electronic Frontier Foundation, which managed to break the algorithm in less than three days at a cost of less than $250,000. The encryption chip that powered the so-called DES Cracker was capable of processing 88 billion keys per second. In addition, it was shown that for a cost of $1 million a dedicated hardware device can be built that can search all possible DES keys in about 3.5 hours.

Since 1998, DES has been replaced with *Triple DES* (3DES), which is based on the DES algorithm, making it very easy to modify existing software to accommodate the new standard. It also has the advantage of proven reliability and a longer key length that eliminates many of the shortcut attacks that can be used to reduce the amount of time it takes to break DES. Even this more powerful version of DES has had a limited life span owing to the increasing processing power of today's computers, and Triple DES is being replaced by AES.

In December 2001, the *National Institute of Standards and Technology* (NIST) announced approval of AES, which specifies the Rijndael symmetric encryption algorithm developed by two Belgian cryptographers. Experts claim that the algorithm is small, fast, and very hard to crack, estimating that it would take 149 trillion years to crack a single 128-bit AES key using today's computers.

The use of AES in wireless devices involves hardware for the processing-intensive encryption process, so older devices may not be upgradeable in many cases. Devices using the AES algorithm still can interoperate with the older devices, but must use the weaker security technologies. AES is optimized for speed and efficiency in the use of memory and system resources, allowing even today's handheld computers to take advantage of it (see Figure 7-2).

Mutual Authentication

Another problem with WEP is that the 802.11b shared-key authentication scheme is one-way, not mutual. An AP authenticates a user, but a user cannot authenticate an AP. Authenticating the AP is important because it is possible for rogue or malicious APs to be on the network (discussed later). For tight security, a mutual authentication scheme between the client and an authentication server must be deployed, forcing both sides to prove their legitimacy within a reasonable time. Because a client and an authentication server communicate through an AP, the AP must support the mutual authentication scheme. This is done with the *Extensible Authentication Protocol* (EAP), which passes messages between wireless client adapters and RADIUS authentication servers.

When the user logs on with a username and password, the client and the RADIUS server perform a mutual authentication with the client authenticated by the supplied username and password. The RADIUS server and

Figure 7-2 Trust Digital offers PDASecure for handheld computers, which supports six different encryption algorithms, including AES, and all are selectable by users. It gives options on which data file to encrypt and which type of encryption to use. PDASecure also prevents unauthorized beaming of data over wireless links.

client then derive a client-specific WEP key to be used by the client for the current logon session. All sensitive information, such as the password, is protected from passive monitoring and other methods of attack. The sequence of events in RADIUS authentication is as follows (see Figure 7-3):

1. A wireless client associates with an AP.
2. The AP blocks all attempts by the client to gain access to network resources until the client logs onto the network.
3. The user of the client supplies a username and password.
4. The wireless client and RADIUS server on the wired LAN perform a mutual authentication through the AP; the RADIUS server sends an authentication challenge to the client. The client uses the user-supplied password to fashion a response to the challenge and sends it to the RADIUS server.
5. Using information from its user database, the RADIUS server creates its own response and compares that to the response from the client; once the RADIUS server authenticates the client, the process repeats in reverse, enabling the client to authenticate the RADIUS server.

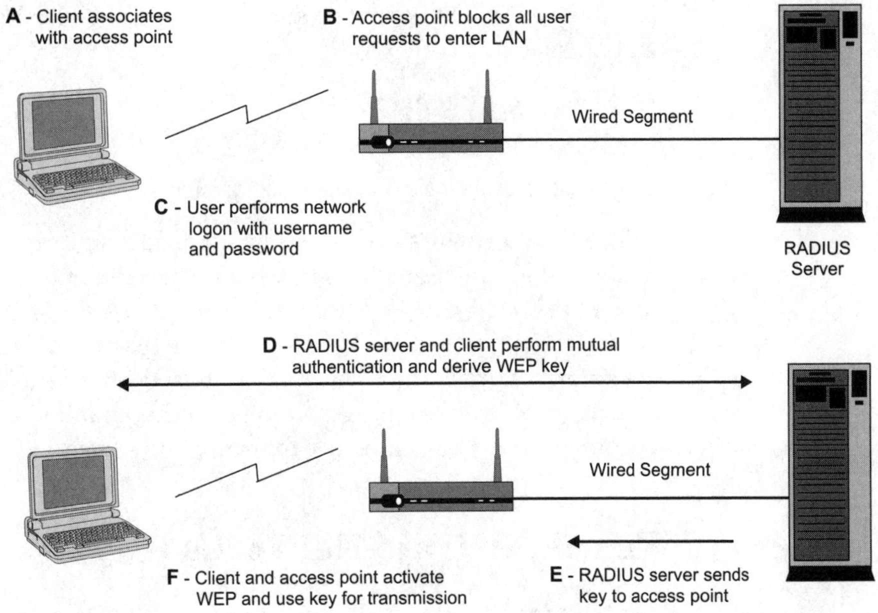

Figure 7-3 The extensible authentication protocol (EAP) provides for two-way authentication between a client and RADIUS server.

6. When mutual authentication is successfully completed, the RADIUS server and the client determine a WEP key that is unique to the client and provides the client with the appropriate level of network access; the client loads this key and prepares to use it for session logon.
7. The RADIUS server sends the WEP key, called a *session key*, over the wired LAN to the AP.
8. The AP encrypts its broadcast key with the session key and sends the encrypted key to the client, which uses the session key to decrypt it.
9. The client and AP activate WEP and use the session and broadcast WEP keys for all communication during the remainder of the session.

The use of per-session encryption keys is important in preventing certain types of hacker activity. Standard WEP supports per-packet encryption but not per-packet authentication. This enables a hacker to reconstruct a data stream from responses to a known data packet. The hacker then can spoof packets for the purpose of gaining access to the network through the AP. By monitoring the 802.11 control and data channels, a hacker can obtain information such as

- Client and AP MAC addresses
- MAC addresses of internal hosts
- The time of association/disassociation

The hacker can use such information to do long-term traffic profiling and analysis that may provide user or device details. One way to prevent this is to ensure that WEP keys are changed frequently, preferably with every client session, as is done by RADIUS. The use of RADIUS has the added advantage of providing a centrally managed, standards-based approach to security that is extensible to wired networks, enabling an enterprise to use the same security architecture for every access method.

Virtual Private Networks (VPNs)

IP-based VPNs are an increasingly popular option for interconnecting corporate locations over the Internet, including branch offices and telecommuters. They also can be used for electronic commerce and making enterprise applications available to customers and strategic partners worldwide. Basically, a VPN lets organizations carve out their own IP-based *wide area networks* (WANs) within a carrier's high-speed Internet backbone. Security functions are performed on IP packets, which are then encapsulated, or tunneled, inside other IP packets for routing across the Internet. By drawing on the economies of transmission and switching that the larger Internet provides, VPNs offer substantial cost savings over private lines or data services like frame relay.

The most widely used protocol for creating VPN tunnels is *IP Security* (IPSec), which ensures the confidentiality, integrity, and authentication of data communication across public networks, such as the Internet.

Confidentiality This refers to the privacy of communications, a fundamental security requirement that is achieved with encryption. The longer the encryption key is employed, the less likely that the privacy of communication will be compromised. Confidentiality is achieved through encryption using Triple DES, which encrypts the data three times with up to three different keys. Although IPSec is used primarily for data confidentiality, extensions to the standard enable user authentication and authorization to occur as part of the IPSec process. This scenario offers a potential solution to the user differentiation problem with wireless connections.

Authenticity Another fundamental security requirement is authenticity. This refers to a mechanism designed to prove the identity of a user prior to the transmission of information. Although IPSec provides excellent protection for moving private data over the public Internet, it lacks an inherent authentication mechanism to verify the identity of the user at the other end of the VPN connection. Establishing this identity can be achieved by applying extensions to IPSec, such as *Internet Key Exchange* (IKE), which authenticates the security gateway or client host at each end of an IPSec tunnel.

IPSec can be applied to secure wireless or wired connections. In either case, an IPSec client is placed on every PC connected to the network and the user is required to establish an IPSec tunnel to route any traffic across the network to a recipient. Filters prevent traffic from reaching any destination other than the VPN gateway and *Dynamic Host Configuration Protocol* (DHCP)/*Domain Name System* (DNS) server.

Integrity This is the assurance that data (either a file or a message) has not been altered during transmission. This is normally achieved by the use of a checksum, which is a simple error-detection scheme. Each transmitted message is accompanied by a numerical value based on the number of set bits in the message. The receiving station then applies the same formula to the message to make sure the accompanying numerical value is the same. If not, the receiver assumes that the message has errors and requests a retransmission.

Role of Firewalls

Deploying a firewall between an AP and a network that only enables authenticated VPNs access can provide additional network protection against attacks. With this method of protection, each client establishes a separate tunnel to the firewall. Authentication of the tunnel is required, which is managed after the network connection is established, requiring additional functionality in either the client or the AP. Alternatively, a VPN gateway can provide this level of protection directly. The VPN gateway can be set up to provide user access rights based on the group the wireless user is associated with.

An organization can tighten security even more by deploying a network-based intrusion-detection system and firewall behind the VPN gateway. That way, before wireless traffic hits the wired network, it will be audited, inspected, and filtered in accordance with the organization's security policy. Building a physically separate infrastructure for wireless access can provide an even higher level of security. Physically separate Layer 2 and 3 segments on dedicated networking hardware can be used to totally isolate the untrusted wireless network until traffic is decrypted at the VPN gateways before being routed into the wired network.

In most situations, using a VPN with a firewall is a good idea whether dialing in from a remote location or connecting via a wireless AP. Adding 128-bit or higher WEP encryption to the wireless session solves the wireless-specific issues.

Certain vendors specialize in providing VPN solutions for the wireless environment. Certicom, for example, offers movianVPN, which brings together the convenience of remote access, the mobility of wireless, and the essential security needed for wireless VPNs. This software client provides end-to-end security from wireless handheld devices to the corporate intranet. It interoperates with popular VPN gateways and supports the IPSec security standard, providing integration with existing wireline VPN infrastructures.

Configuring the movianVPN client entails the IT administrator defining the MAC and IP addresses of the remote gateway. The information requested next depends on which gateway the organization uses. In general, a user ID and password (preshared key), as well as the IP address or subnet on the remote network, will be entered as part of the configuration process. The IPSec options will also be configured, including the encryption scheme. The client software comes with several diagnostic and troubleshooting tools, such as a ping utility and an IKE log to troubleshoot IPSec VPN errors.

Even if the organization is not using a VPN, a firewall is essential for guarding the corporate LAN. One of the most basic mistakes many organizations make when they deploy wireless networking is to treat the wireless component as if it were simply an extension of their existing wired LAN. This means that should hackers gain access to an AP, they have full access to all the company's internal systems, applications, and storage. Therefore, rather than allowing APs to be connected directly to the network, all network ports that have an AP connected should be routed through the firewall. At the firewall, all policies that are applied to Internet connections to guard against intrusions can simply be adjusted to include wireless network access.

Proprietary Security Solutions

Because WEP has proven to be so weak, vendors have taken it upon themselves to offer their own security solutions until a more robust standard emerges from the IEEE. The drawback to these solutions is that the organization must commit to a single vendor for all of its wireless equipment.

Psion Teklogix, for example, offers its 802.IQ protocol, which replaces the *Transmission Control Protocol* (TCP)/IP on the wireless network. The published methods of breaking WEP depend on knowing the structure of the radio data packets, assuming they are TCP/IP packets. Drawing on the protocol-filtering capabilities of an AP enables all other protocols to be left off the wireless network except for the proprietary 802.IQ packets, and this effectively closes the system to all but Psion Teklogix mobile computers. Even if the packets are decrypted, the structure of the 802.IQ packets is not published so the data remains unreadable.

Based on the 802.IQ architecture, the data on the wireless network passes between the wireless network controller and the mobile client only, which reduces the amount of data that is exposed to eavesdropping and eliminates the possibility of an active intrusion. Eavesdropping on such sensitive transaction information as *Stock Keeping Unit* (SKU) numbers and quantities used for placing orders, when taken out of context, holds little value to hackers.

Several other vendors provide extra security in ways that require that the client and AP come from the same vendor. The previously described EAP is Cisco's solution that uses a proprietary algorithm to support mutual authentication between a client and a RADIUS server. The RADIUS server must support the EAP algorithm for this to work. Another vendor, Agere, provides a solution that uses a nonstandard dialog between the client and AP to enable standard *Point-to-Point Protocol* (PPP) authentication methods to be used. This allows any standard RADIUS server to be used to authenticate a user. 3Com provides a way to create tunnels using the *Point-to-Point Tunneling Protocol* (PPTP) and uses standard RADIUS to authenticate the tunnel.

Kerberos has been discussed within the IEEE as a possible security solution for wireless networks, but this is a client-server solution that may be too burdensome for mere APs to handle. Accordingly, Symbol Technologies supports an authentication server that implements Kerberos as part of its mobile security architecture. The advantages of Kerberos include the following:

- It provides all levels of standard security services, including confidentiality, authentication, integrity, access control, and availability.
- It ensures roaming between APs, resulting in uninterrupted application connectivity.
- It enables the reauthentication of a client to the network in less than 40 milliseconds.
- It provides a standards-based security service supported by Windows platforms and Linux distributions.
- Although Kerberos can be used for mutual authentication and key generation, the downside of Kerberos implementations is that they require an IP address for the requestor, and in the wireless framework the IP address is often assigned after authentication, rather than before.

Addressing Security Threats

Wireless networks are vulnerable to a number of security breaches. Hacker tools are even available for free download over the Internet to help anyone break into a wireless network. Nevertheless, IT administrators can minimize the exposure to risk in several ways.

Hardware Theft

It is common to assign a static WEP key to a client that is stored either on the client's disk storage or in the memory of the client's WLAN adapter. When this is done, anyone in possession of the client has access to the MAC address and WEP key and can use them to gain access to the WLAN, even after it is lost or stolen. It is virtually impossible for an IT administrator to detect a security breach that results from misuse of the MAC address and WEP key until the owner reports it. Then the administrator must change the security scheme to render the MAC address and WEP key useless for WLAN access and the decryption of transmitted data. This involves changing the encryption keys on all clients that use the same keys as the lost or stolen client. The greater the number of clients, the larger the reprogramming task.

To deal with the ramifications of lost or stolen clients, a more stringent security scheme must be implemented. Requiring a device-independent username and password for authentication is a good start. This not only grants users access to the network regardless of the clients they may be using at any given time, but since they are device independent, the username and password are not stored in the client where hackers can find them. And instead of using static WEP keys that are also stored in the client, WEP keys should be generated dynamically on user authentication.

Rogue and Malicious APs

The shared-key authentication scheme specified in the 802.11b standard for WiFi networks provides for one-way, not mutual, authentication. In other words, the AP authenticates a client, but the client has no way of authenticating an AP. It is simply assumed that the AP is legitimate and has a right to be on the network. This makes it easy for employees to bring in their own APs and plug them into the network through the wall jack in their cubicle to enjoy the benefits of mobility. As a result, the APs are behind the corporate firewall, which opens up a huge hole in security of which the IT department may be totally unaware.

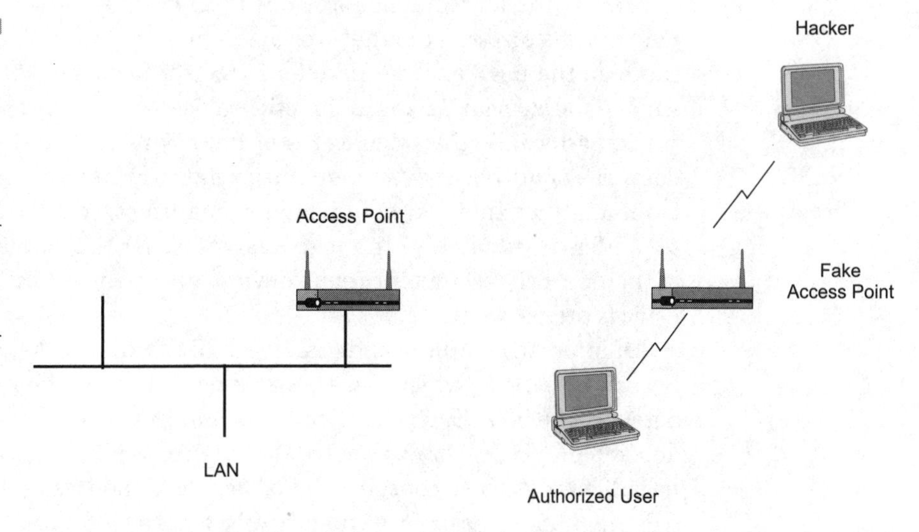

Figure 7-4 A man-in-the-middle attack relies on a fake AP being installed near a legitimate AP to trick the user into revealing his or her username and password.

A related problem is the possibility that an AP may be secretly plugged into the corporate LAN for purposes of industrial espionage. This can be done by individuals who have unfettered access to the building, such as maintenance, housekeeping, and security people. All it takes is for the AP to be plugged into any RJ45 wall jack in an unoccupied office or cubicle via *Category 5* (CAT5) cable. The devices are so small that they can easily be hidden from view and may go undiscovered for weeks or months while the intruder snoops from afar. The goal of a rogue is to get valid traffic off the network. Once the rogue AP is in place, the hacker can intercept traffic with a man-in-the-middle attack.

To implement a man-in-the-middle attack, the fake AP is positioned near a legitimate AP (see Figure 7-4). A legitimate user connects to the fake AP in error and enters his or her username and password. The hacker collects this information and informs the user that the username and/or password is incorrect. After several unsuccessful attempts to log on, the user gives up in frustration. After a few minutes, the hacker uses the legitimate user's username and password to log onto the real AP.

One way to detect the presence of rogue or malicious APs is to send IT staff through the building with WLAN-equipped laptops and sniffer software. When the client comes into range of an AP, the AP's beacon signal is picked up, which triggers an audible tone from the client, indicating the presence of the signal. But this can be time-consuming and costly, especially in big buildings or a multibuilding campus. Products that automate this task are now available on the market.

AirDefense, for example, provides a suite of tools that makes it easy to pinpoint the electronic signatures of the majority of WLAN APs and access cards on the market. The AirDefense toolset includes both a walk-around sniffer and special APs that function as sensors that can detect 802.11b transmissions, so that signatures of unknown APs can be compared to a database of authorized gear. Another vendor, Finisar, offers a WLAN spectrum analyzer that can help pinpoint unauthorized APs. Network Associates offers Sniffer, a tool that can analyze either the wireless or wired side of the network to pinpoint rogue devices and measure activities the rogue devices are being used for.

An interesting approach is taken by IBM with its Distributed Wireless Security Auditor, which uses authorized wireless clients as sensors to detect rogue APs. Each client runs a small Linux program that sniffs and detects all APs, reporting their IP and MAC addresses to a central database. That database contains the IP and MAC addresses of all authorized APs, making it easy to automatically determine whether a device is a rogue. The auditor package also includes triangulation software, which enables IT administrators to pinpoint the physical location of unauthorized APs. The tool could be scaled to monitor large networks from a central point, such as the WLANs used in hundreds of facilities operated by a multinational corporation.

Hiding APs

Software tools can be downloaded for free from the Internet, such as Network Stumbler and Kismet, which automatically search the airwaves for vulnerable APs. The practice of traveling around a city searching for unprotected APs is known as *war driving*. However, software can be used to hide APs by generating tens of thousands of phony AP beacon frames, thus hiding legitimate APs from hackers.

One product, called Fake AP, aims to thwart would-be hackers by generating 53,000 bogus wireless APs in the vicinity of a real AP. Only those clients that know the true identity of the AP will be able to quickly distinguish the real point of entry through this airwave camouflage. Although more knowledgeable individuals will spot these bogus APs for what they are, newbie hackers will be effectively deterred. This software does not provide a foolproof security solution, however, because experienced hackers can write programs to quickly test every beacon frame, similar to the way they have written programs to quickly dial phone numbers or test passwords.

Hardening the WLAN

IT departments can take a number of steps to harden WLANs against potential attacks. These steps do not require detailed knowledge of security; in fact, with only the technical expertise to get wireless connections running, these procedures can be easily implemented.

Check for Same-Channel Conflicts

Typically, vendors ship all their equipment with the same default channel. When users set up the wireless link, they might not think of changing the default channel and are especially reluctant to do so after getting the wireless link running, fearing they will mess something up and never get the link back. But if other networks are operating in the area and they are also using the default channel, they could all inflict *denial of service* (DoS) attacks on each other through radio interference.

To avoid this situation, the configuration software provided by the manufacturer of the AP or client adapter should be used to look for other networks that may be within range. If other wireless networks are operating in the area, the channels they are using will be displayed. To avoid any interference, it is just a matter of changing the default channel with the configuration software to one that is as far away as possible from those already in use nearby.

Change All Default Settings

It is recommended that all default settings on all network components be changed. Default information for all WiFi vendors is widely available on the Internet in newsgroups, bulletin boards, and on manufacturer web sites. Tools such as APSniff (see Figure 7-5) and Network Stumbler (see Fig-

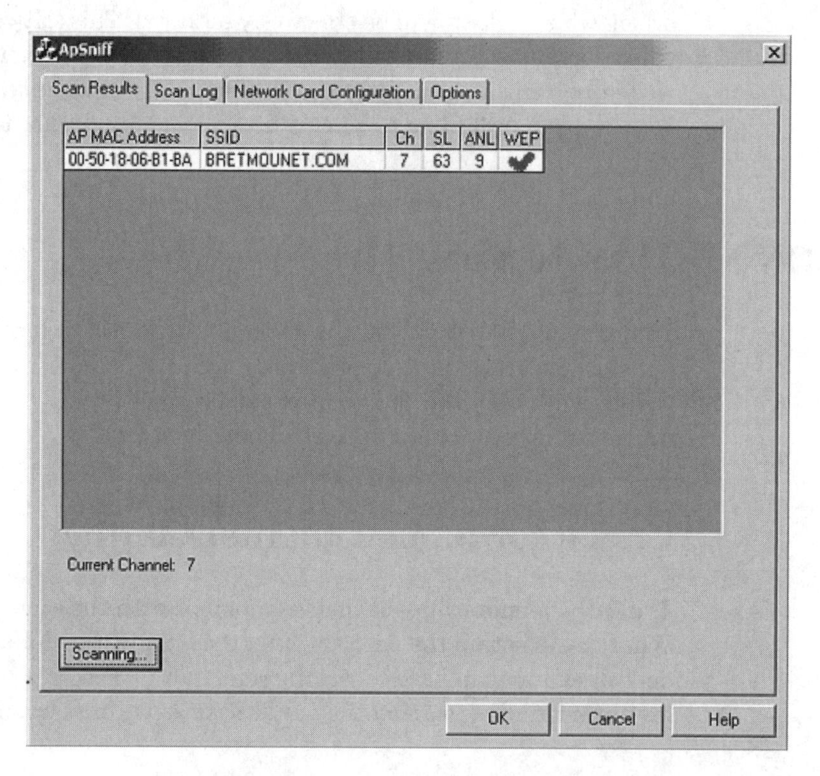

Figure 7-5 APSniff is a Windows-based tool used to discover APs that are broadcasting beacons. Its legitimate uses include identifying interfering APs and helping set up wireless clients by providing client configuration information.

ure 7-6) enable hackers to see critical network settings in a WiFi network, even if WEP has been applied.

If the default settings are still in place for the WiFi network, and it is unprotected by WEP, then it is likely that the other defaults for other components may be in place as well. For instance, the router default password or possibly access to network shares may be open.

Apply WEP

Despite its flaws, WEP still provides protection against snooping coworkers and casual hackers. What requires some effort, however, is key management. This refers to the generation, distribution, updating, and revoking of the keys used to encrypt and/or digitally sign information. The trouble with any security system that uses encryption keys is that the keys are susceptible to compromise by hackers or through disclosure. Good key management addresses these issues in a variety of ways, including

- Changing the keys at specific intervals
- Protecting the manner in which keys are distributed
- Publishing the certificate revocation lists of keys known to be compromised or expired so they will not be accidentally used

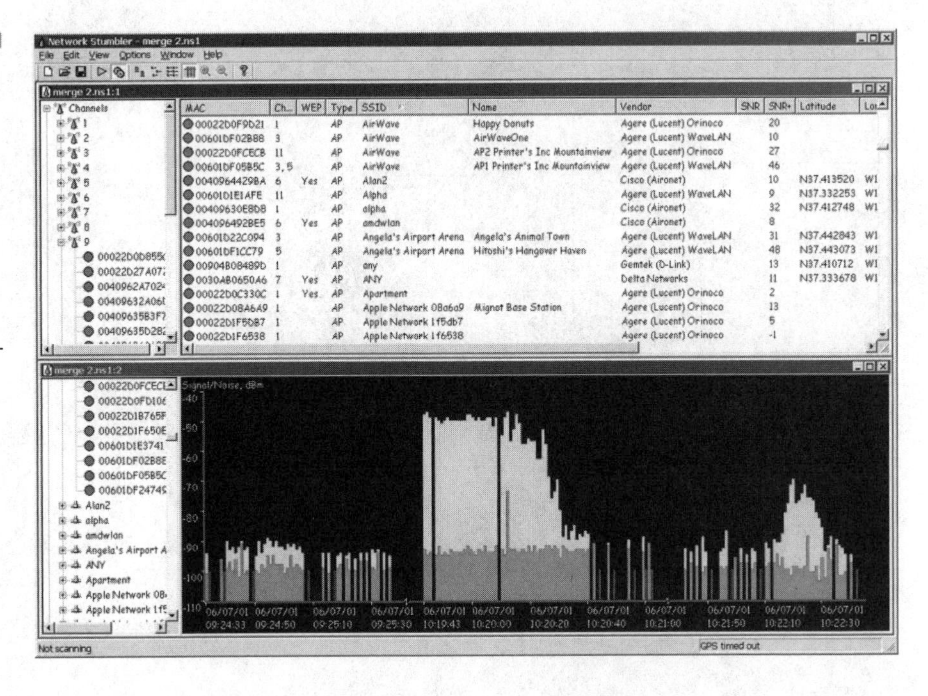

Figure 7-6 Network Stumbler is a Windows-based wireless AP sniffer that enables the user to obtain a list of all APs broadcasting beacon signals.

Even if only one copy of a WEP key is found or captured, the entire system could be compromised. The original WEP specification issued in 1997 supported unique keys for each station, but this is rarely implemented; instead, a single key is normally created for all users and this key is normally never changed. Similarly, no prescribed distribution mechanism is used; administrators have been known to e-mail keys in the clear to other users, and since no controls have been put in place around key management, they will likely never know that a key has been disclosed. The same applies to attacks via crypto-analysis tools used by hackers. If a key has been cracked and it is never changed, the intruder will have free access until it does get changed.

Use a Firewall

The AP should be considered an untrusted network, just like the Internet, with all the same threats. Therefore, the AP should be placed within a *demilitarized zone* (DMZ) in front of a firewall (see Figure 7-7).

Figure 7-7 AP network placement in relation to the internal corporate network

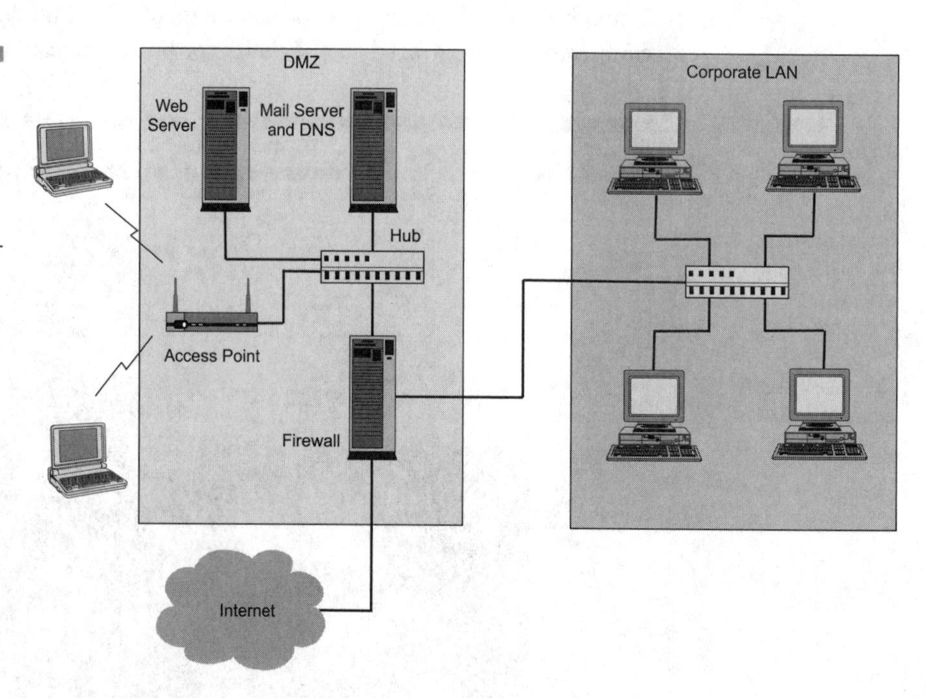

If the organization does not have a central firewall, each computer on the internal corporate network could be inexpensively equipped with client software that provides firewall functionality. This type of product is available from vendors such as BlackICE, McAfee, Symantec, and Zone Labs. These client firewalls automatically detect and block attacks through a comprehensive inspection of all inbound and outbound information. They can be used to secure dial-up, *Digital Subscriber Line* (DSL), and cable modem connections, as well as wireless links.

Implement MAC Filtering

Most WiFi products support MAC filtering, which can be used to harden the wireless network against intrusion. The MAC address is a 12-character code that is unique to each piece of hardware on the network. The addresses are assigned by the manufacturer, and it is possible to use these addresses to restrict access to the network. The MAC address can be viewed through the client interface software that is installed with the 802.11 hardware. An AP can be configured to enable connections from a set of client devices identified by their MAC addresses. If a device attempts to connect to the AP but its MAC address is not in the access list, entry to the network will be denied.

Access lists of MAC addresses do not provide hackerproof security, however, simply because MAC addresses can be forged. Software is available on the Internet that enables anyone to define a MAC address for any given device. Some WiFi products even offer this capability in their client or AP configuration tools, enabling hackers to configure their equipment with a stolen MAC address. If a hacker can listen in on a wireless network long enough, the MAC addresses can be learned. When a legitimate client device logs off the network, the hacker can log on with a stolen address. The AP will have no way of knowing one device from another, especially if WEP is not turned on.

An improvement on MAC address filtering involves the implementation of RADIUS, which can be used to manage a MAC access list for multiple APs and update this information on a scheduled basis. This saves the administrator from having to configure each AP with the same MAC-permission information and try to keep that information updated in a timely manner.

Administrators can use additional tools to keep an eye on network activity in real time. One of these tools, freely available on the Internet, is *arpwatch*, with arp being the acronym for Address Resolution Protocol.

Originally developed for wired Ethernets, this tool can be set up to monitor wireless links as well. The function of arpwatch is to watch MAC addresses as they enter and leave the network. If an unknown MAC addresses is detected, an alert will be triggered and the administrator will be notified via e-mail. Another thing arpwatch finds is duplicate IP addresses. When it finds a duplicate, it logs the old and the new MAC addresses. The tool will even detect a MAC address spoof, which occurs when a legitimate IP and MAC address pair suddenly changes to a new MAC address, indicating that a hack is in progress. The report messages generated by arpwatch include the following:

- **New activity** This MAC/IP address pair has been used for the first time in six months or more.
- **New station** This MAC address has not been seen before.
- **Flip flop** The MAC address has changed from the most recently seen address to the second most recently seen address.
- **Changed MAC address** The host (that is, client) switched to a new MAC address.

Disable DHCP

If a wireless router has been deployed, assign static IP addresses to the wireless client adapters if possible and turn off DHCP. This creates more administrative overhead, but it also prevents the DHCP server from passing IP addresses to any client trying to associate with an AP. Although a wireless sniffer could easily pick out IP addresses, turning off DHCP makes the wireless network tougher to access by the casual drive-by intruder.

Change the Subnet

If a wireless router is deployed and DHCP has been disabled, the IP subnet should also be changed. This is recommended because many wireless routers default to the 192.168.1.0 network and use 192.168.1.1 as the default router IP address. Even if a network does not give out an IP address, a knowledgeable hacker can assume that the router is using the default setting. The hacker can just configure his or her notebook with an IP address in the 192.168.1.0 network using 192.168.1.1 as the router address, and the hacker can obtain access to the Internet through that network.

Restrict Beacons and Probe Responses

APs advertise their availability and the configuration parameters they support by means of beacons. In large wireless deployments, these beacons enable users to distinguish one AP from another. If the configuration of the AP changes, users can still find it through the beacon. Through the beacon, users can roam into a wireless environment and receive notification that the service is available without having to ask an administrator.

The beacon interval is set by the manufacturer and can be reset or even shut off by the system administrator. Shutting off or disabling the beacon prevents configuration information such as the SSID, channel, rate, and WEP from being broadcast to all devices within range. This stops essential information required to associate with an AP from being picked up by any and all listeners, including hackers.

Another type of signal is the probe request, which is issued by client devices looking for APs when they arrive in the wireless environment between beacons. This signal is broadcast on a select channel and all APs within range, by default, will respond with a probe response, which essentially contains the same information as the beacon. War-driving tools discover WLANs by broadcasting probes on all channels with the expectation of receiving responses from APs. When a response is received, the configuration information that was returned by the AP is input into the hacker's equipment configuration utility.

Although it is not realistic to turn off probe responses because it would disable the entire wireless network, it is possible to modify the configuration of the AP so that it only responds to probe requests with the SSID specified. In this way, the required association parameters become far less easy for hackers to obtain by simply listening to the channels and broadcasting probe requests.

Monitor Traffic Volume and Set Limits

Many times intruders will generate a large amount of WLAN traffic, especially if their intent is to capture corporate data. In such cases, they will download whatever they can find and sort through it at a more convenient time. Intruders may also simply be looking for free, high-capacity network access. Whatever the intent of the intruder, the MAC address will have a significant amount of data flowing to it. By monitoring the amount of data going to a device, administrators can zero in on suspected intruders for a closer examination.

System administrators may also want to apply global limits on traffic volumes. This would discourage intruders from trying to sell wireless access to the Internet through a company's network. Depending on the manufacturer, it might be possible to implement storm threshold filtering at the AP to minimize this kind of bandwidth theft. With this feature, limits can be set on the number of packets per second from a specific MAC address or a total volume of data on a given port on a specific interface. Once it is determined that a storm is occurring, any additional packets from that MAC address will be denied. The storm will be determined to be over after 30 seconds has elapsed where every one-second period has less than the stated threshold in broadcast packets.

Select Appropriate Broadcast Strength

Most APs and other wireless devices come with the antenna set to maximum broadcast power. This default setting maximizes the range of the wireless link and minimizes the requirement for the manufacturer's technical support when users call in to complain of weak signals. But it is often the case that more broadcast power is being used than is actually required for a given wireless link. This extends the range of the wireless link, leaving it exposed to war-driving hackers.

A typical AP uses a pair of dipole antennas for diversity. One or the other will be used, depending on which gets the stronger signal. However, these antennas also extend the range of the signal far beyond what is actually needed for local client access, enabling intruders to intercept data. System administrators can minimize this RF leakage in the following ways:

- If possible, do not place APs against exterior walls or near windows. Plan signal coverage to radiate only as far as the windows, but not beyond. Many APs are sold with an omnidirectional antenna, which means that a wireless signal is radiated equally in all directions. Place one near a window and it will transmit as far into the street as it will in the office.
- Centralize the APs as close to the center of the usage area as possible to increase signal strength in the service area and reduce radiation leakage.
- Select locations with office furniture and interior walls to dampen the signal and further reduce external leakage.

- It may be possible to set the antenna power level on the AP. If so, reduce the power of the antenna gradually while testing for signal strength.

In the last step, the objective is to bring the power level to the lowest point while still providing coverage for good data throughput and reception. The primary advantage of this technique is that the AP is likely to remain hidden from nearby eavesdroppers, since they are less likely to find the wireless link while driving at street level.

Be aware, however, that this is not always a dependable security measure. If a hacker already knows of a wireless link's existence, the use of a high-gain antenna can retrieve the signal from a location that is out of range for normal devices. Also, people one floor above and below will still be able to pick up the wireless signal, leaving open the possibility of an inside attack.

Control the Radiation Profile

An effective way to secure the wireless link is to simply make it unavailable to those who have no need for it. After all, if the signal cannot be received by a device, it cannot be compromised or disrupted. Some APs and PC Cards come with external antenna connector ports, which, when used, override the internal antenna. These ports make it possible to implement specific signal radiation profiles.

The radiation pattern can be flattened and shortened in ways that minimize wireless signal leakage into nonsecure areas where intruders might pick it up. Not only can the radiation profile be adjusted manually, but the configuration tools that come with the external antenna can be used to adjust the strength of the signal to prevent radiation from extending beyond the required range.

Use Directional Antennas

In high-density environments, it is not unusual to have multiple wireless networks competing for the available spectrum and inadvertently denying each other service. Other times, an unauthorized device may be causing interference. These problems can be diagnosed by using a directional antenna that can lead administrators to the offending device.

Affordable kits are available that include device-tracking and spectrum analysis software as well as a high-gain directional antenna for identifying rogue APs and other devices. Similar functionality can be approximated using any 2.4 GHz directional antenna and a portable 802.11 device with an antenna interface, but identifying the location of a specific device will be more difficult without the spectrum analysis software.

One of the most portable tools of this kind comes from *Berkeley Varitronics Systems* (BVS), which offers its YellowJacket testing system designed to operate on a Compaq iPAQ so users can analyze 802.11b WLANs. A specially designed radio unit slides underneath the iPAQ and the YellowJacket software uncovers such information as AP identification, the packet error rate, SSID, and multipath and received signal strength (see Figure 7-8). A spectrum analyzer display indicates actual signal levels across the entire 2.4 GHz band (see Figure 7-9). Signal levels can be monitored in real time and captured in a cumulative mode to better understand the environment over time. This enables sources of RF interference and other WLANs operating nearby to be accurately characterized. The tool also provides direction finding for locating individual APs and stations.

Bird's Eye mapping software from BVS enables the creation of facility floor plans and the plotting of WLAN information on-the-fly as the administrator or technician walks through a facility. The software displays AP MAC addresses, received signal strength, and SSIDs directly onto the floor plan. It also depicts RF coverage, which is helpful when performing RF site surveys and rechecking the coverage of an operating WLAN.

Conduct Preemptive Scans

Administrators should survey corporate facilities periodically using a tool like Network Stumbler to see whether any rogue APs show up. The signal strength outside the building should also be checked. Even if the signal is weak and throughput is only 1 or 2 Mbps, it is still a security breach that must be corrected; otherwise, all other security measures become irrelevant.

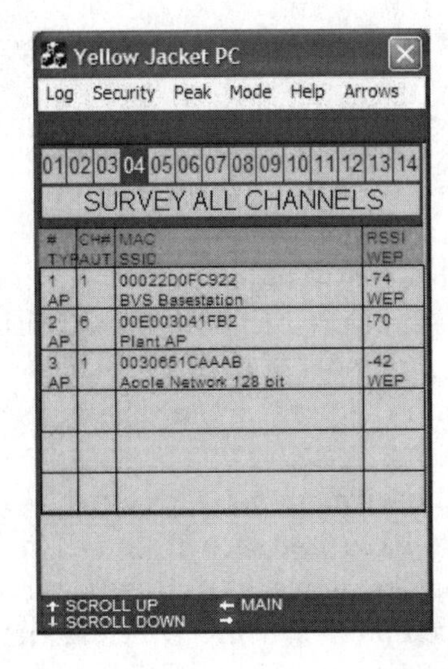

Figure 7-8 BVS offers the YellowJacket testing system for handheld and portable computers, which scans all WiFi channels to identify active devices.

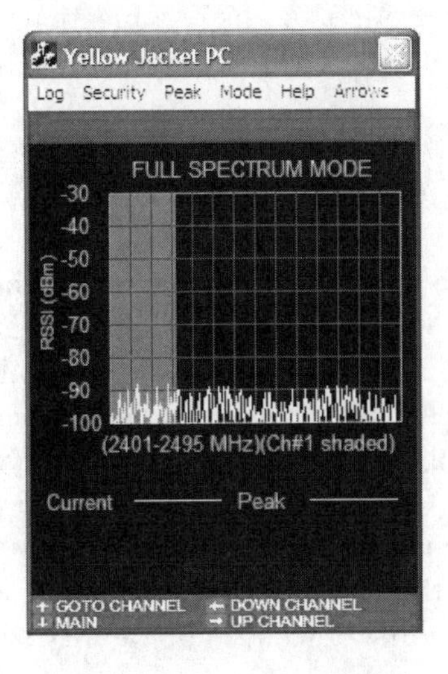

Figure 7-9 The YellowJacket testing system from BVS includes a spectrum analyzer display, which indicates actual signal levels across the entire 2.4 GHz band.

Conclusion

No WiFi product provides an all-encompassing security solution. Each vendor's security offering addresses only a piece of the problem and, worse, security solutions from different vendors are not interoperable. Even when vendors claim to adhere to security standards, they seem to have a proprietary component that makes their products unable to work with other vendor's offerings. Consequently, when security is concerned, an enterprise should expect to be locked into the wireless cards and the APs of one vendor for the foreseeable future.

Maintaining security for the wireless network must be an ongoing activity. Hackers are relentless in exposing security holes, and wireless links make it easier to breach corporate networks. Industry surveys indicate that as many as 80 percent of companies using wireless networks have not taken even the most basic security measures. This makes for easy pickings for hackers. When conducting an attack against a wireless network, hackers do not even feel the need to cover their tracks; they simply use an unprotected network, which becomes the source of an attack on another company's network. By the time anyone traces the source of the attack, unsuspecting IT managers are left to answer questions from authorities, since all signs point to their network as the source of the attack. Meanwhile, the hacker is long gone.

Moreover, the company whose network was identified as the source of the attack may face a downstream liability lawsuit if the hacker used it as a launchpad to cause harm to another business. After all, the victimized company might seek to recover its losses and try to convince a court to award damages, especially if it could be proven that the network owner failed to exercise reasonable care in securing its systems. The failure to take the most basic security precautions could even be construed as negligence. Although no legal precedent for such lawsuits exists yet, the *Computer Emergency Response Team* (CERT) Coordination Center has warned of the possibility that companies could be held liable if their networks are used in attacks. As if to underscore this possibility, some insurance companies now offer policies that protect against downstream liability lawsuits.

CHAPTER 8

Troubleshooting the WLAN

Wireless local area networks (WLANs) occasionally experience connectivity problems that must be dealt with by administrators and technicians. Common problems in a wireless network have to do with throughput, configuration, interference, and the cable segments. Some problems go beyond what site survey tools can detect before installation, including changes in the operating environment. Such changes would consist of moved walls due to employee growth, security breaches, and improper configuration of equipment.

Although a *Wireless Fidelity* (WiFi) network may easily satisfy a handful of users when initially set up, as the rest of the organization starts seeing the benefits of wireless, more users and applications are likely to be added to the WLAN. This increases network utilization, which often leads to poor throughput that annoys users. Without the use of effective support tools, the administrator may not know that a problem exists until users start complaining to the help desk.

This situation is made worse by the difficulty in forecasting the performance of an 802.11b wireless network to begin with. The 11 Mbps data rate of 802.11b does not mean that each user can obtain that rate for sending or receiving information; at best, the actual throughput is closer to 5 or 6 Mbps. As the number of users increases, throughput will diminish further. If some of the users are transferring bulky files, the performance of the wireless link deteriorates even more. Roaming users will strain the wireless link as well. Therefore, for performance and security reasons, administrators must monitor traffic on the wireless network to head off performance problems.

Other times, the problems with a WiFi network are the result of improper equipment configuration. Despite their *graphical user interfaces* (GUIs), the configuration utilities offered by vendors are often confusing and may result in setting the wrong *service set identifier* (SSID), having network usage conflicts with other users because of hidden nodes, and conflicting *access point* (AP) channel allocations. In some cases, such as if a user has an SSID that does not match the one at the AP, the problem results in no network connection at all. Hidden node problems and channel allocations often result in poor performance. Many factors can be the source of poor performance, and without the proper tools and techniques, it can be difficult to track down the fundamental cause.

The Role of Protocol Analyzers

As with wired LANs, a protocol analyzer is an indispensable tool for wireless networks. With special software that runs on a notebook or *personal digital assistant* (PDA) equipped with an 802.11b radio card, the cause of many problems can be traced to the source. The key functions performed by a wireless-equipped protocol analyzer include

- **Monitoring and capturing** The analyzer passively monitors and captures all data traffic sent over the WLAN within radio range of the unit and stores the data on the laptop's hard drive. The passive operation of the analyzer avoids adding any performance impact on the network. The analyzer can also monitor signal strength and display it for each station based on minimum, maximum, and current values.
- **Decoding** The analyzer opens the received data and decodes its protocol type and frame structure (see Figure 8-1). This includes the definition of all 802.11 control, management, and data frames, as well as other information embedded within the frames, such as SSIDs, AP channels, and the data rates of each user's radio, showing how many packets are sent and at what speed: 1, 2, 5.5, or 11 Mbps. Most analyzers also decode higher-layer protocols, such as AppleTalk, *Internetwork Packet Exchange* (IPX), NetBEUI, and *Transmission Control Protocol/Internet Protocol* (TCP/IP). If configured with the proper *Wired Equivalent Privacy* (WEP) key, the analyzer can even decode encrypted data.
- **Filtering** The analyzer filters the received data and then performs statistical analysis, displaying the results in various ways. This offers a flexible means for monitoring and troubleshooting the network. For example, an analyzer can display all AP and user SSIDs, which will help spot whether an end user is trying to associate with an AP using the wrong SSID. The analyzer can also be set up to display the actual throughput of all users by specific channels and identify unauthorized users and APs that do not fit an acceptable profile. Depending on the specific analyzer used, information about all the channels can be displayed in a single consolidated view.

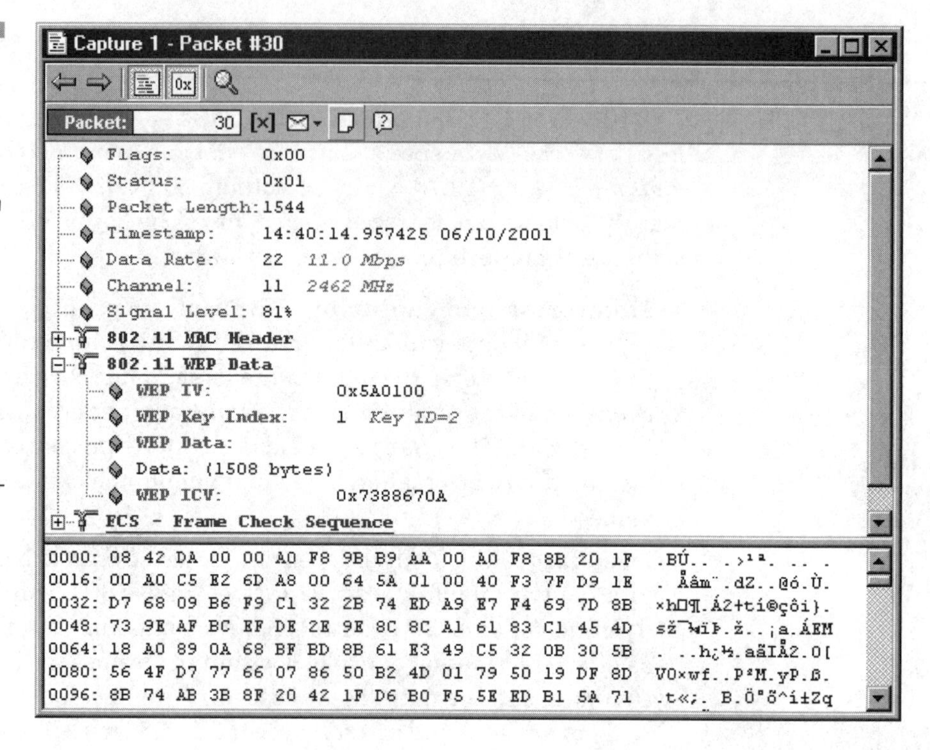

Figure 8-1 The single-packet decode feature of AiroPeek from WildPackets, Inc. works on 802.11a and 802.11b protocols, displaying management, control, and data packets, as well as all higher-level network protocols.

The use of a protocol analyzer can spot security risks in real time, identify network problems efficiently, and reduce network operating costs. An existing protocol analyzer used for wired LANs can be equipped with an 802.11b (2.4 GHz) or 802.11a (5 GHz) radio card and software to function in the wireless environment.

Network Instruments offers 802.11b and 802.11a wireless support in its Observer line of network monitor and protocol analyzers. In addition to wireless troubleshooting tools, Observer offers a number of security and intrusion-detection facilities, including the ability to alert administrators when unauthorized users try to gain access to the WLAN, improper WEP usage, and exposure of rogue APs. The fact that these features, full wireless and wired troubleshooting, are combined into one unit makes it easier to isolate problems.

Among the capabilities of Observer is that it identifies errors on the WLAN by station (see Figure 8-2) and aggregates signal measurements. It measures wireless speeds as well as signal strength by station. It also takes AP statistics (see Figure 8-3). In AP statistics mode, Observer shows traffic passing through any AP, displaying the following

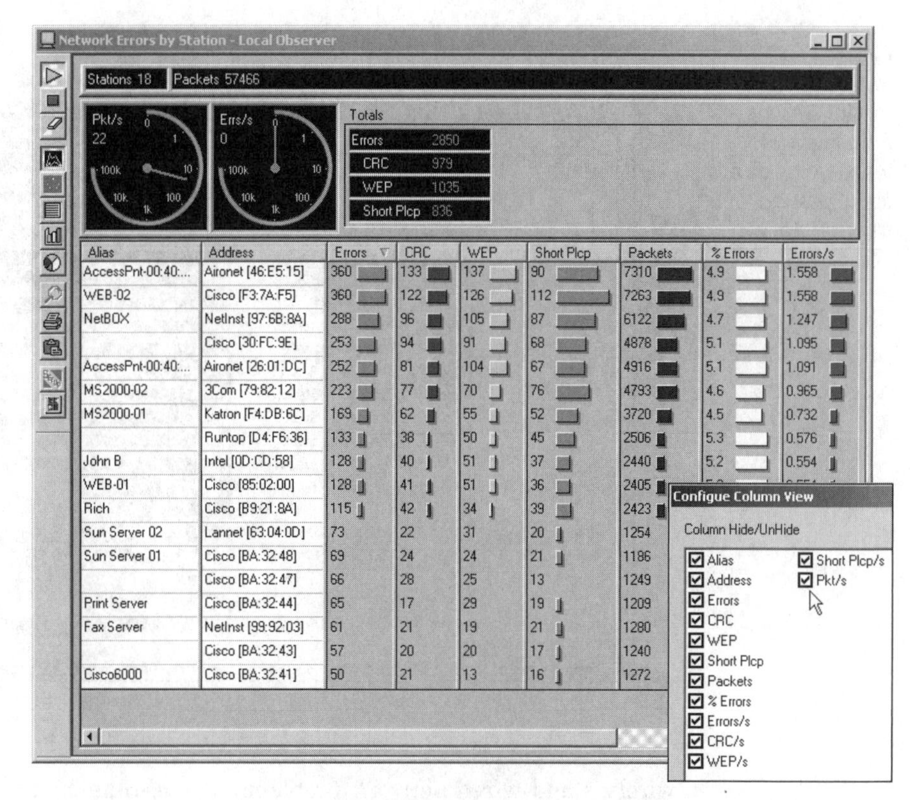

Figure 8-2 Network Instruments' Observer identifies wireless errors by station, enabling administrators to pinpoint errors as they happen. Errors displayed include cumulative and rated Cyclic Redundancy Check (CRC), short Packet Layer Convergence Protocol (PLCP), and WEP.

- Wireless stations that are connected to an AP
- Nonwired stations the AP communicates with
- Levels of signal strength, quality, data transfer rates, and nondata transfer rates on each station on the AP
- AP traffic totals

For example, an administrator can immediately see if a station has connected to the wrong AP or if an unauthorized AP has been installed. AP statistics will display whether a station has a problem with quality or the range of the connection based on the number of reassociations and retransmissions, or whether a station has been misconfigured based on station poll totals.

Wireless triggers and alarms detect errors and intrusions. A trending capability enables wireless statistics to be viewed over days, weeks, months, and years for baselining performance. The trending capability provides a record of performance over time for each network segment. Performance can be compared to the original baseline measurement to determine the amount of deviation. A wireless analysis capability identifies existing and potential

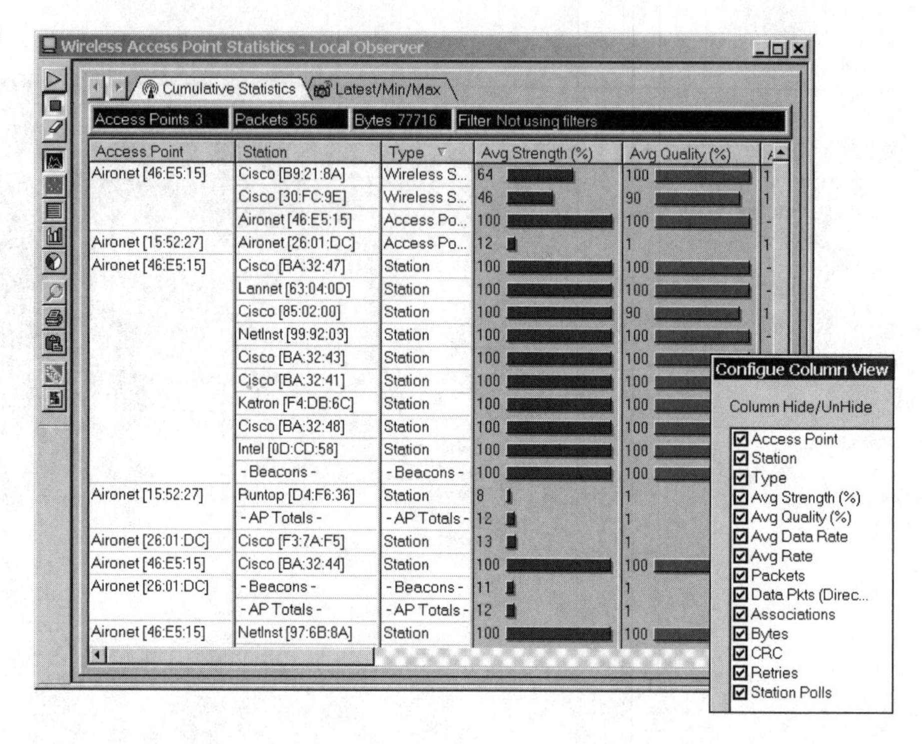

Figure 8-3 In AP statistics mode, Observer from Network Instruments shows traffic passing through any AP.

wireless and wired network problems in real-time or after traffic capture. Observer's expert notification system can identify over 400 kinds of events.

Basic Connectivity Issues

After installing wireless client adapters and APs, things may not work as expected, especially since quite a lot of detailed information is required by the manufacturer's configuration utility. Furthermore, wireless technology is still not well understood by administrators and technicians who have spent most of their careers tending wired networks.

IP Address Assignment

If no connectivity exists after setting up client adapters, APs, bridges, or routers, one of the first things that should be done is a simple ping test. This is done using the operating system's command line or a third-party ping utility. A ping is done using the target device's IP address. A successful ping will return several replies. If the ping times out, this indicates that something is wrong at the target device that prevents the ping packets from reaching it.

The next step would be to check the IP address of the target device, making sure that it is entered in the correct format, that it does not duplicate an existing IP address, and that the device is on the same subnet. For example, if the IP address of an AP is 192.168.0.50 with a mask of 255.255.255.0, the client adapter's IP address must be in the same address range, 192.168.0.x, and have the same mask of 255.255.255.0.

If the IP address of a client device is not known, perhaps because a *Dynamic Host Configuration Protocol* (DHCP) server may assign it at network logon, the plain language name of the device can be used for the ping. If the ping is successful, it will also return the currently used IP address of the device (see Figure 8-4).

If the IP address of a target device is known but the *Media Access Control* (MAC) address is not, the *Address Resolution Protocol* (arp) command can be used with the -a switch to discover it on the network. The format of this command is

```
arp -a 192.168.100.5
```

The results are displayed in Figure 8-5. The MAC address might be needed for entry into a protocol analyzer to monitor the busiest transmitting nodes on the local network and display test results by source-destination address pairs. Test results include the frame count transmitted by each

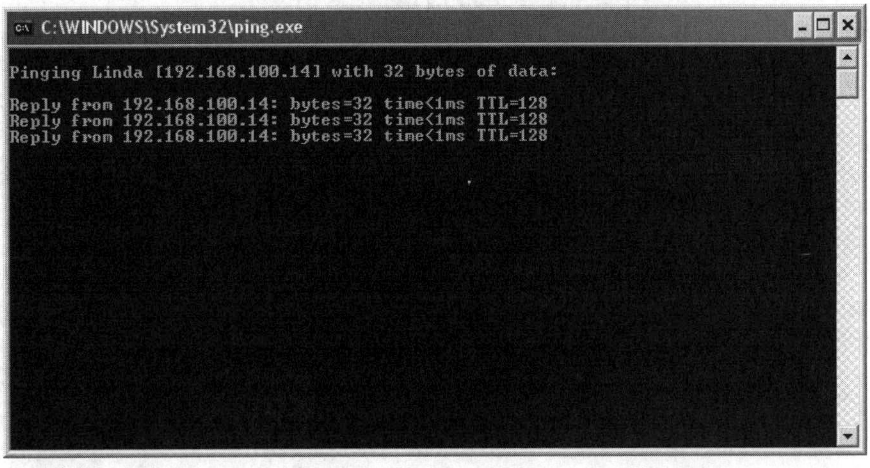

```
C:\WINDOWS\System32\ping.exe

Pinging Linda [192.168.100.14] with 32 bytes of data:

Reply from 192.168.100.14: bytes=32 time<1ms TTL=128
Reply from 192.168.100.14: bytes=32 time<1ms TTL=128
Reply from 192.168.100.14: bytes=32 time<1ms TTL=128
```

Figure 8-4
A successful ping using the plain language device name Linda. The ping also returns the target device's current IP address of 192.168.100.14.

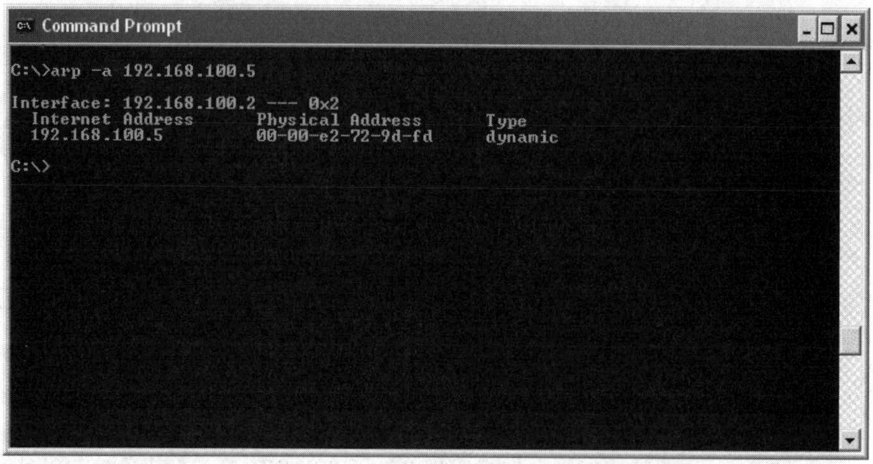

```
C:\>arp -a 192.168.100.5

Interface: 192.168.100.2 --- 0x2
  Internet Address      Physical Address      Type
  192.168.100.5         00-00-e2-72-9d-fd     dynamic

C:\>
```

Figure 8-5 Results of the arp command with the -a switch, showing the MAC address of the target device associated with a particular IP address

address in the pair combination. When address filtering is turned on, the protocol analyzer can display conversations to or from the filtered address. Some protocol analyzers can even identify the manufacturer of the client adapter by the first three bytes of the MAC address.

Cabling

APs are cabled to the corporate LAN via a hub or switch. If intermittent connectivity problems or an excessive number of errors that bog down throughput occur, the cable length may be greater than the recommended Ethernet segment length. With *Category 5* (CAT 5) cable for 10/100 Mbps segments, the length should not exceed 328 feet. If the distance from the hub or switch exceeds the recommended segment length, using a fiber or a wireless hop, such as a repeater, can rectify the problem.

Interference is another problem that can affect the performance of cable. It occurs when network cables are run near high-power equipment, especially when installed in warehouses and factories where heavy machinery may be in use. Fiber is recommended in such environments because of its immunity to noise, and since fiber does not radiate signals, it is also more secure than copper media or wireless connections. Although fiber costs more than CAT 5 cable, its advantages make it worthwhile for harsh environments. Most APs do not come equipped with fiber ports, so a media converter may be required.

Radio Interference

A site survey should be conducted before installing a wireless network and should be done on the actual site under normal operating conditions with all wireless devices present. Such a survey is important because *radio frequency* (RF) behavior varies with the physical properties of the site and cannot be predicted accurately without doing a site survey. For example, construction materials like steel and wood absorb RF energy to varying degrees, as do objects with water content. The presence of microwave ovens, cordless phones, and the latest generation of fluorescent lighting can also interfere with 2.4 GHz signals, as can the RF signals from nearby WiFi networks.

Interference may be continuous or intermittent. Continuous interference is noticed when the link is first put into operation, but intermittent interference can occur unexpectedly at any time, such as when a microwave oven in the employee snack room is suddenly turned on. The remedy for this situation might be as simple as moving the WiFi device out of range of the microwave oven. Over a larger distance, however, it may be difficult to determine if interference is even present and to what extent it may be causing performance problems.

The first step is to find out if the receive signal level is adequate at both ends; this measurement will reflect both the desired and undesired (interference) signals being received. If the *Received Signal Strength Indicator* (RSSI) is correct and interference is suspected, the *Signal-to-Interference Noise Ratio* (SINR) should then be measured at each end of the system. This parameter and the received signal strength can be plotted and tracked as a function of time by using the radio histogram feature provided with the manufacturer's site-planning tool. The two parameters can then be tracked simultaneously and correlated with observations of the degraded performance.

For example, if the received signal strength remains consistently good and the SINR exhibits periods of abnormally poor readings, the link performance (the error rate and so on) should be checked during the periods of poor SINR. If a correlation takes place, this confirms that the link is indeed being affected by interference. When the RSSI is adequate at both ends, interference will usually be intermittent.

The most effective way to diagnose interference is to use a spectrum analyzer. The analyzer shows a visual image of the frequency band of interest, highlighting interfering signals, if they exist. The toughest cases involve sources of interference that are turned on and off at irregular intervals. For these cases, it is imperative to have a spectrum analyzer that can capture and store its images over time, so that briefly appearing sources of interference can be discovered without the need to sit in front of the analyzer for hours.

Intermittent connectivity problems in certain locations and during certain weather conditions might point to the need for a site survey, if one was not done prior to initial installation. An essential element of the site survey consists of checking the signal strength at various locations where APs or bridges will be placed (see Figure 8-6). The equipment should be placed in locations where no sources of interference occur and where the most users will be served while minimizing signal leakage into uncontrolled areas where intruders might pick them up.

Antennas

If it is confirmed that interference is a problem and no steps can be taken internally to eliminate it, a visual survey should be done at both ends of the wireless link to see if any other antennas or related types of equipment are

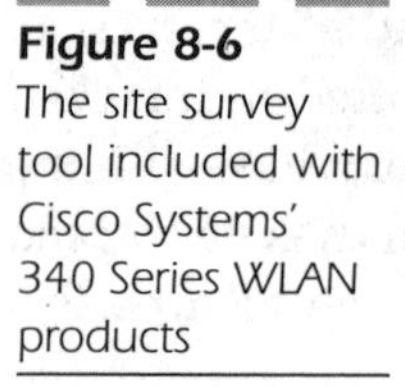

Figure 8-6 The site survey tool included with Cisco Systems' 340 Series WLAN products

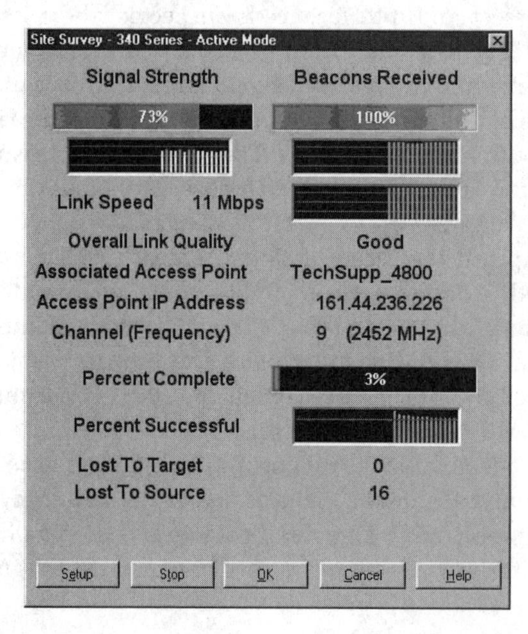

in the area. If other radio equipment is seen, try to determine who owns or operates it so that additional information can be obtained, such as

- What frequency or frequencies does the system use?
- How much transmit power does the system use?
- What type of antenna polarization is being employed?
- What are the hours of operation?

Upon contacting the owner/operator, find out if they would be willing to assist in finding out if their equipment is indeed the source of the interference problem. Start with the antennas:

- Are one or more antennas pointed at each other?
- Can the antennas be repositioned to eliminate an intersecting path?
- Can the polarization of antennas at one location be changed so that they are cross-polarized?

If the problem persists, the frequency of one of the systems may have to be changed. An easy way to change the frequency is to simply swap the transmit and receive frequencies using the system's configuration utility. Most of the time, changing frequencies will be effective, but since WiFi systems operate in the unlicensed bands, different user organizations must cooperate in working out interference problems. The *Federal Communications Commission* (FCC) and the courts typically will not get involved in such issues except in cases where a law has been violated.

Another way to deal with an interference problem is by changing antenna gain. Higher gain results in a narrower beam width, which can make the RF system less susceptible to a source of interference. Because this solution might involve changing antennas, it is the most costly and time-consuming solution.

Client Adapters

WiFi client adapters come in several form factors: *Peripheral Component Interconnection* (PCI) cards for desktop computers, notebooks, and PDAs, and even Compact Flash for PDAs and other handheld devices. WiFi components also come embedded in notebooks and other portable devices. When a problem arises, it is usually with the add-in cards, rather than the components that come already embedded in notebooks and PDAs, which are thoroughly tested by the manufacturer before shipment.

Indicator Light-Emitting Diodes (LEDs)

The easiest way to check for problems with the client adapter is to view the *light-emitting diodes* (LEDs) on the card (see Table 8-1). The client adapter will have two LEDs:

- **Link integrity/power LED (green)** This LED lights when the client adapter is receiving power and blinks slowly when the adapter is linked with the network.
- **Link activity LED (amber)** This LED blinks when the client adapter is receiving or transmitting data and blinks quickly to indicate an error condition.

Resource Conflict

A possible problem with newly installed adapters of any type is resource conflict, and WiFi cards are no exception. If the client adapter does not communicate after driver installation, it may be that a conflict exists with other devices. Although the more recent editions of Windows minimize the possi-

Table 8-1

Summary of conditions indicated by various LED combinations

Green LED	Amber LED	Condition
Off	Off	Either the client adapter is not receiving power or an error has occurred.
Flashing quickly	Flashing quickly	Power is on, the self-test is successful, and the client adapter is looking for a network.
Flashing slowly	Flashing quickly	The client adapter has associated with an AP.
Continuously on or flashing slowly	Flashing	The client adapter is transmitting or receiving data with an associated AP.
Off	Flashing quickly	The client adapter is in the power-save mode.
On	Flashing quickly	The client adapter is in ad hoc mode.
Off	On	The driver is incorrectly installed.
Off	Flashing in a pattern	An error has occurred.

bility of conflicts due to their support of the *Plug-and-Play* (PnP) specification, occasionally it still happens. When it does, an *Interrupt Request* (IRQ) can be selected for the card that is not in use by another device.

Years ago, the IRQ settings were configurable via *dual inline package* (DIP) switches on the card. Today this is done within Windows. With Windows XP, for example, possible conflicts can be identified using Device Manager, which displays devices that share the same resources, such as *direct memory access* (DMA) channels, *input/output* (I/O) ports, IRQs, and memory addresses. To open Device Manager, click Start and then Control Panel. Next, double-click System. On the Hardware tab, click Device Manager. If an IRQ must be changed, the following procedure is used within Device Manager:

1. Double-click the type of device that requires a change.
2. If the device has resource settings that can be changed, the Resources tab will appear in the dialog box.
3. Click the Resources tab, and then clear the use automatic settings check box. (This check box might be unavailable if a non-PnP device is selected.)
4. In Setting based on, click the hardware configuration to change.
5. In Resource settings, under Resource type, click the resource type to change: DMA, IRQ, I/O port, or memory address.
6. Click Change Setting, and then type a new value for the resource type.

When installing a client adapter on a computer running Windows NT, the operating system will prompt the user to confirm or modify the factory-set device values. The Windows NT diagnostics program, however, is not bulletproof. Occasionally, a hardware conflict will occur despite the fact that the Windows NT diagnostics program has displayed resources as being available for the wireless client adapter. This may be the case when the computer has one or more devices and/or peripherals installed that claimed an I/O base address and/or an IRQ value without notifying the Windows NT operating system. In this case, the Windows NT diagnostics program will not be able to display these values simply because it is not aware they are already in use.

If this is the case, the conflict may result in a wireless client adapter that does not work and/or a conflicting device that does not work properly. It will depend on the user's preferences and the configuration options supported by each of the conflicting devices to determine which device settings should be changed to enable flawless operation.

Verifying Communication

It may be necessary to verify that the client adapter can communicate with an AP. This is done by checking the AP's association table through the management system to see if the client adapter is listed. The management system's association table page lists all the devices, both wireless and wired to the root LAN, of which the AP is aware. The association table lists information about each device, such as its name, IP and MAC addresses, state (associated or authenticated/unauthenticated), and parent.

However, if the client adapter has associated with an AP but is unable to use the network, the Ethernet side should be checked to determine if the AP is communicating properly with the LAN. From a computer on the wired LAN, launch a ping at the AP using its IP address. If the ping is not successful, the problem could be a bad cable or connector. If the ping is successful, then the problem could be a bad driver on the client adapter.

Other Considerations

If the client adapter seems to be working but the user is unable to connect to the network, the problem might be due to a configuration mismatch. For example, if both LEDs on the client adapter flash once every 10 seconds or so, the problem is likely to be caused by a configuration mismatch of either the network name or the encryption key.

With some vendors' products, the network name is case sensitive, and with MS-DOS systems, the network name should be entered between quotation marks. If the quotation marks are omitted, the driver will interpret the value as all uppercase, as in 15FACE instead of "15Face." With some products, if the APs have been configured with both lower- and uppercase characters, the driver will not be able to establish a radio connection.

If a network connection cannot be established but the LEDs on the client adapter indicate that the network is working properly (that is, the power LED is on and the transmit-receive LED flashes), check that the TCP/IP settings of the network are set. This is especially important when the client is running an older version of the Windows 95 operating system, which does not automatically install the necessary TCP/IP stack. The solution to this problem is to get out the Windows 95 disk and configure the computer for networking. During this process, the TCP/IP stack can be selected for installation.

If the wireless client adapter has been installed successfully but it does not become operational, it is possible that an I/O port conflict is taking place. The Windows Resource Manager may not have been able to detect the conflict, allowing the installation to complete successfully, but not allowing the client adapter to operate properly because another device has been assigned the specified port. The solution would be to change the default I/O base address used by the wireless client adapter.

Access Points (APs)

The client adapter is not always the source of connectivity problems; sometimes the AP needs tweaking to get the wireless link up and running properly, especially since the same unit may serve as an AP, repeater, or bridge, all of which may be installed on the same wireless infrastructure. This means ample opportunity exists for configuration errors.

Root Mode

An important factor in the proper operation of an AP is the root mode setting; it must be set based on how the AP interacts with other devices in the wireless infrastructure. An AP will be configured as a root device if it accepts associations with and communicates only with clients and repeaters. Although it can be one of many root devices per wireless network, the AP must not be set up to communicate with other root devices.

An AP configured as a nonroot or repeater device, however, associates and communicates to a root or another nonroot device associated to a root. It accepts associations and communicates only with clients and repeaters, as long as it is registered to a root device.

IP Addressing

Another source of common installation problems relates to IP addressing. For example, without the *Trivial File Transfer Protocol* (TFTP) server IP address, it is impossible to download an image file to the AP. The image file is the AP's executable program. If this program gets corrupted or the admin password is forgotten, a new image can be downloaded from the TFTP server to restore the AP to proper operation.

Downloading a new AP image is done over an Ethernet connection to the computer the TFTP server resides on, which can be any computer on the

LAN. Before starting the image transfer, the administrator needs to know the AP's IP address, subnet mask, the TFTP server IP address, and the AP image file name. The TFTP server must be configured to point to the folder containing the image to be downloaded. Upon rebooting the AP and the administrator performing a few simple procedures, the image file will be automatically downloaded to the unit. When downloading is completed, the AP is ready for configuration, providing it has a correct IP address.

LED Messages

The operating status of an AP can be readily determined by its external indicator lights, which have different meanings that may vary by vendor. On the Aironet 340 series AP from Cisco Systems, for example, the indicators have the following meanings:

- The Ethernet indicator flashes green when packets are received or transmitted over the Ethernet infrastructure.
- The status indicator flashes green when the AP is operating normally but is not associated with any wireless devices. Steady green indicates that the AP is associated with a wireless client. (For repeaters, the status indicator flashes differently, according to its association or nonassociation with a root AP.)
- The radio indicator flashes green to indicate radio traffic activity.

As an aid in troubleshooting problems, it is important to become familiar with the messages indicated by the LEDs. For example, if the status indicator flashes green, no client devices are associated; therefore, the unit's SSID and WEP settings should be checked as possible causes of this problem. (Wireless clients that attempt to associate with the AP must use the same SSID as the AP.) If the Ethernet indicator is blinking red, the Ethernet cable is disconnected. On the other hand, if all three indicators (Ethernet, status, and radio) are displaying steady red, this may point to a firmware failure, in which case simply disconnecting power from the unit and reapplying it may correct the problem.

Security

If security is enabled on one station on the network, it must be selected on all stations. Likewise, if security is disabled on one station, it must be deselected on all stations.

If WEP is turned on to protect communication over the wireless link, the key used to transmit data must be set up exactly the same on the AP and on any wireless devices with which it is associated. For example, if WEP key 2 is set on a wireless client adapter to 0987654321 as the transmit key, WEP key 2 on the AP must be set to the same value. However, the AP does not need to use the same key for its transmissions as long as it and the target client both support the same WEP keys. If one or the other does not have a matching WEP key, data traffic cannot be passed because the data is encrypted.

During troubleshooting, it is advisable to disable WEP until the connectivity problem is identified. After things are working properly, WEP can be enabled. If the connectivity problem returns, the cause has just been narrowed to the WEP configuration in one or more devices. Another factor that bears attention is the fact that some products permit WEP keys to be written in either ASCII or hexadecimal format. Sometimes the ASCII format proves troublesome. To eliminate this as even a potential problem, all WEP keys should be written in hexadecimal.

Recovering Default Settings

If the administrator forgets the password that enables configuration of the AP, the device will have to be completely reset to return all AP settings to the factory defaults. The settings that will be returned to their factory defaults include passwords, WEP keys, IP addresses, and SSIDs.

Other Issues

Many AP products can be found on the market today. Although they include a standard set of capabilities and features, making them interoperable with other vendors' products when used on the same network, they may be also equipped with a number of proprietary features designed to differentiate them in the increasingly crowded market for WiFi products. Accordingly, some troubleshooting routines may be associated with these features.

DHCP Server Some APs can be configured as a DHCP server, and more than one AP on the network can function as a DHCP server, providing that the address pool of each server does not overlap. Care should be taken to prevent any AP configured as a DHCP server from conflicting with any other DHCP server on the network. Difficulty may be experienced if the pool size of the DHCP server is less than 8 when using wireless stations that run Microsoft Windows. This is because Windows sometimes requests several addresses before accepting an offered IP address. This can cause several addresses to be temporarily unavailable for distribution to other wireless stations when they are also logging onto the network.

Turbo Mode Some vendors' WiFi products offer an optional turbo mode feature that enables the data transfer speeds of 802.11b equipment to burst from 11 to 22 Mbps and 802.11a equipment to burst from 54 to 72 Mbps. During the installation, the turbo mode setting (enable/disable) must be exactly the same for the AP and wireless clients. The wireless connection will not be established in either infrastructure or ad hoc mode if the turbo feature is enabled on one end of the wireless connection and not the other. If the AP and wireless client adapters come from different manufacturers, it is not guaranteed that the turbo feature will be interoperable between them. Instead, they will likely default to the standard 802.11b or 802.11a data transfer speeds of 11 Mbps and 54 Mbps, respectively.

Bridges

Bridges extend LAN segments for a greater range or can be used to break larger networks into more manageable subnets. The bridge must be within radio range of the AP and both devices must have the same SSID. Likewise, the WEP key used to transmit data must be set up exactly the same on both the AP and the bridge.

Root Mode

Only one bridge can have the setting of Root On in a wireless network; all other bridges should be set to Root Off. A bridge with Root On accepts association and communicates only with clients and repeaters. It will also not communicate with other Root On devices. A device set for Root Off associates and communicates with a Root On or Master bridge. It can also associate and communicate with clients and repeaters as long as it is registered to a Master bridge.

LED Messages

The operating status of a bridge can be readily determined by its external indicator lights, which have different meanings that may vary by vendor. For example, if the radio and status indicators flash amber and steady green, respectively, the radio's maximum number of retries or buffer thresholds may have been reached. This may mean that the AP the bridge is communicating with is overloaded or radio reception may be poor. The administrator might be able to correct these conditions by changing the bridge's SSID to communicate with another AP or by repositioning the bridge to improve connectivity.

Line of Sight

For long-distance communications, consideration must be given to the Fresnel zone in addition to the line of sight. The line of sight, as its name implies, is the straight line between two points, or the visual path. The Fresnel zone is the elliptical area immediately surrounding the visual path. This area varies depending on the distance the signal must travel and the frequency of the signal. If the elliptical area contains obstructions, reliable communication will not be possible, even if the visual path is unobstructed.

The time to take the Fresnel zone into account is during site planning, but over time, new obstructions may be introduced into the environment, such as the construction of a new building or the growth of vegetation. The Fresnel effect is overcome by raising the antenna height at one or both locations. Distance calculation spreadsheets are available from equipment vendors, which provide the height of the antenna for the given radio distance.

Spanning Tree Protocol (STP)

The *Spanning Tree Protocol* (STP) is used to remove loops from a bridged LAN environment. The protocol also enables the bridges in an arbitrarily connected infrastructure to discover a topology that is loop free (a tree) and ensure that a path exists between every pair of LANs (a spanning tree). However, it is possible for a bridge to be blocked by STP; a leased line or an alternate path might be set up between the points bridged in the RF network, and STP might have put one of the links in the blocking mode to avoid loops.

Problem Diagnosis

Most of the discussion so far has centered on problems related to installation. Diagnostic tools can be used to solve problems with installed systems as well. Real-time activity counts at each point in the system enable technicians to determine if data is being processed properly at each point along the path to its destination. Such tools also maintain counts on several types of transmission errors and the stage at which they occur.

With a bridge, for example, the unit has an Ethernet port and a wireless port. Traffic statistics are collected as the data passes through these ports and may be used to diagnose performance problems at the management system's GUI. The following statistics are collected as data arrives at the bridge's LAN port:

- *Rx frames* provides a count of frames received from the LAN at the bridge's Ethernet port. To determine if packets are lost at any processing stage, this count is compared to corresponding counters at other processing stages, such as Tx Frames, Rx Dframes, and Tx Dframes. Although lost packets indicate a problem, other indicators must be looked at to isolate the cause.
- *Rx alignment* provides the number of misaligned frames received. If this counter increases rapidly, it might be due to heavy traffic on the local Ethernet from many collisions, bad cable connections, or bad adapters. For the wireless port on the bridge, this could be due to RF noise, a low signal level, or bad radio modules. Normally, this counter should display 0 or remain static.
- *Rx CRC errors* provides a count of CRC errors. If this counter increases rapidly, as with the Rx alignment counter, it might be due to heavy traffic on the local Ethernet from many collisions, bad cable connections, or bad adapters. For the wireless port on the bridge, this could be due to RF noise, a low signal level, or bad radio modules. Normally, this counter should display 0 or remain static.
- *Rx no buffers* provides the number of frames lost due to a lack of frame buffers in the bridge. When this counter increases rapidly, the system is most likely overloaded. Normally, this counter should display 0 or remain static.
- *Rx bad packet* provides the number of partial or incomplete frames received. When this counter increases rapidly, the incoming packets

may be damaged. This could be due to an Ethernet equipment problem or the bridge becoming overloaded. Normally, this counter should display 0 or remain static.

Another set of counters collects information on frames being sent out of the bridge's the wireless port:

- *Tx Dframes* provides the number of frames sent out of the wireless port. This count can be compared to corresponding counters at other processing points, such as Rx Frames, Rx Dframes, and Tx frames.
- *MAC state* provides an indication of the MAC state if the bridges are unable to hear each other for any reason. Several link states can be reported:
  - 1 = Link down
  - 2 = Sending
  - 3 = Link up (idle)
  - 4 = Receiving
  - 5 = Data ack (substation only)
- *Tx no resources* provides a count of frames lost due to the lack of buffer space. If this problem is persistent, it may require a reconfiguration of the network.
- *Tx frames dropped* provides the total number of frames dropped due to the lack of an RF link. Frames also can be dropped due to an overload in the wireless network.
- *Tx pool or percent buffer available* provides the percentage of available transmit buffer space. With an overload condition, 0 percent indicates that no resources are available.

Another set of counters collects information on frames received at the wireless port of the bridge:

- *Rx Dframes* provides a count of frames arriving at the wireless port of the bridge. This count can be compared to corresponding counters at other processing points, such as Tx Frames, Rx Frames, and Tx Dframes.
- *Signal level* is a measure of signal strength. A signal level exceeding 30 percent is considered a strong signal. The stronger the level of noise along the path, however, the stronger the signal must be to overcome the noise.

- *Signal-to-noise ratio/signal level* indicates the strength of the received signal relative to the local noise. This measurement is updated each second. If the *signal-to-noise ratio* (SNR) is poor and the signal level is high, the antenna may be too close to a local noise source. If SNR is poor and the signal level is low, the problem may be signal attenuation caused by obstructions or the antennas being placed too far apart.
- *Rx CRC errors* provides the number of frames lost due to CRC errors. A high volume of CRC errors can indicate a problem, but other counters must be viewed to isolate the cause. Normally, this counter should display 0 or remain static.
- *Rx alignment* provides the number of misaligned frames received. When this count increases rapidly, it may indicate an RF signal problem, the presence of interference, or noise in the signal. These problems can be solved by trying another RF channel or running a diagnostics utility to determine the cause. Normally, this counter should display 0 or remain static.
- *Rx bad packet* provides the number of partial or incomplete frames received. Occasional increases in this count are normal, but when the count increases rapidly, it indicates the packets may be damaged because of a weak signal or interference. The presence of ambient noise may cause such problems. If the environment contains noise, choosing another RF channel may solve the problem or running a diagnostics utility may determine the cause. Normally, this counter should display 0 or remain static.
- *Rx other network ID* provides the number of frames received on any network ID other than the current unit. When this count increases at more than 10 per second, the bridge is receiving and rejecting signals from a nearby competing RF system. In this situation, another RF channel should be tried. Normally, this counter should display 0 or remain static.
- *Rx timeout* provides the number of times a station failed to reply to the base station within the timeout period. Generally, this indicates a failure in the wireless system. Normally, this counter should display 0 or remain static.
- *Tx frames* provides a count of frames sent by the Ethernet port. This count is compared to corresponding counters at other processing points, such as Rx Dframes, Rx Frames, and Tx Dframes.
- *Tx no resources* provides the number of frames discarded because of a lack of buffers. This generally indicates an overloaded local LAN.

- *Tx frames dropped* provides the total number of frames dropped due to errors such as excess collisions.
- *Tx pool* indicates available transmit buffers. A value of 0 would cause the Tx no resources counter to increment.

The Role of Management Systems

Turnkey operational solutions enable administrators to manage and troubleshoot problems with the wireless infrastructure. Cisco, for example, offers its CiscoWorks *Wireless LAN Solution Engine* (WLSE), which provides template-based configuration based on user-defined groups to effectively manage a large number of Cisco Aironet wireless network APs and bridges. It provides critical monitoring information, including the status of the link to the authentication server, congestion at each AP, and connectivity between the AP and the network. It also enhances general security management by detecting improperly configured APs and bridges, ensuring the uniform deployment of wireless access policies to multiple devices through its configuration templates.

An appliance-like device, WLSE provides administrators with a single console for configuring, troubleshooting, and maintaining wireless APs. It also proactively monitors the wireless infrastructure by simulating key events and generating notifications if services are unavailable or show performance deterioration. Reports are provided that aid in capacity planning by identifying the most active APs. These reports help determine if APs should be relocated or added while shortening the time needed to solve client access problems.

Prior to WLSE, Cisco's APs could only be maintained through embedded management systems in each device. Although an embedded approach is serviceable for environments with only a few APs, a centralized approach is more cost and time efficient as wireless networks grow. Although no firm rule exists for determining when an organization should step up to a centralized solution, when the network grows beyond 50 devices it starts to become beneficial. When the network grows to more than 100 devices, administrators will want a centralized system just to save time.

The primary benefit of a centralized approach is for an enterprise to carry out configurations, maintenance, and repairs automatically, eliminating the need to work on a device-by-device basis. APs, for example, rely on almost constant updates to firmware in order to keep track of the latest technology and the most recent changes to standards. In addition to constant updates, wireless APs often require frequent configuration changes to reflect modifications in access and security policies.

Without a centralized approach, a technician would be required to separately configure each AP in the network, a process that could take as long as 30 minutes. With a centralized management system, an administrator or technician could simply create a single configuration profile and automatically download it to all the APs in the network.

Computer Associates is among the network management vendors that have added wireless extensions to their products. Unicenter is the company's flagship management platform. It has been enhanced with modules that enable corporations to manage all 802.11b usage on an enterprise-wide scale. The products, Unicenter *Wireless Network Management* (WNM) and Unicenter *Mobile Device Management* (MDM), split the management of 802.11 networking between APs and mobile devices. They also treat 802.11 networks as one part of the networking whole, which means that from a management viewpoint little difference exists between a traditional wired network and a new 802.11 wireless network.

Together, WNM and MDM automate the management of 802.11 wireless networks while adding capabilities not found in the average wireless network. WNM, for example, has an advanced discovery capability that automatically finds APs and the devices that are connected to them. With this information, WNM can then capture events on the devices to help administer quantify performance and availability. In addition, WNM offers enhanced security, including intrusion detection, using existing security tools built into Unicenter.

Since the product is built on top of Unicenter, the number of APs that can be managed has few limits. In theory, hundreds of thousands of networks can be managed using WNM. Because of variances between APs, however, WNM will not automatically support every 802.11b and 802.11a AP on the market. This is due to the 802.11 specifications not yet providing a uniform management interface.

Unicenter MDM casts a wider net, supporting both 802.11 devices and non-802.11 devices. MDM focuses on such aspects of device management as software delivery, asset management, and VPN management. In addition, as most mobile devices do not have antivirus support, MDM will monitor potential virus issues on these devices, including viruses that are not native to the device but that can be passed on to a system where the virus can take root.

Conclusion

With their ease of installation and portability, wireless networks are gaining popularity and creating new network troubleshooting issues for companies. In addition to the standard capacity and access issues, administrators now face security and expandability concerns as well as the challenge of integrating wireless technology into existing LAN infrastructures. Almost without exception, new wireless networks are connected to wire networks. Until recently, wireless analysis offerings were limited, both in the type of troubleshooting information offered and in their lack of integration with traditional protocol analysis tools for wired networks. Now tools offer wireless capabilities to ease the increasingly complex nature of network administrators' jobs.

CHAPTER 9

Network Management

Enterprise applications and mission-critical data that are essential to business operations rely on the availability and continued reliability of the corporate network. But the introduction of wireless links to extend the corporate network adds complexity to network management, which requires the appropriate tools to manage, monitor, and troubleshoot the network.

The purpose of a network management system is to improve network availability (up time) and service reliability, centralize the control of network components, reduce staff time for administrative tasks, and contain operational and maintenance costs. The network management system can effectively lower the cost and complexity of today's networks by providing a set of integrated tools that enable IT staff to quickly isolate and diagnose network problems. The ability to analyze and resolve network problems from a central location is critical to the management of both network and personnel resources. The minimum functional requirements of any network management system are as follows:

- **Fault management** This function includes the detection, isolation, and correction of events that are responsible for abnormal network operation. Fault management provides the means to receive alarms, determine the cause of a network fault, isolate the fault, and implement corrective action.
- **Configuration management** This function is used for the setup, maintenance, and updating of network components. Configuration management also includes the notification to network users of pending and completed configuration changes.
- **Accounting management** This tracks network usage to detect inefficiencies, the abuse of network privileges, or unusual network activity, all of which are useful for planning network changes or growth.
- **Performance management** This recognizes current or impending performance issues that can cause problems for network users. Activities include the collection and analysis of statistics for determining baseline network performance, monitoring, and maintenance to ensure acceptable network performance.
- **Security management** This includes the activities of controlling and monitoring access to the network and associated network management information. It also includes controlling passwords and user authorization as well as collecting and analyzing security or access logs.

Ideally, these and other functions should be available from the same network management system, which treats the entire network as a single entity. In practice, however, network management systems vary in sophistication, forcing organizations to rely on different tools from different vendors to obtain all the functionality they need to properly manage their networks. Increasingly, corporate programmers are even applying their UNIX script-writing skills to enhance freely available open-source management utilities. Within a relatively short time, they can create management tools that meet their wireless environment needs simply and economically instead of having to work with limited *application programming interfaces* (APIs) or petition a vendor for desired features, which may or may not show up in a subsequent release.

Management System Evolution

Network management systems started out by providing simple monitoring and alerting functions that helped to ensure that network links were operating properly. Monitoring simply determined whether a link was up or down; alerting involved displaying or printing a statement that the link was up or down, or sending a notification to a technician via e-mail. Devices were monitored and controlled by element managers. Each vendor had his or her own element manager for its products. In general, network and element managers relied on *Simple Network Management Protocol* (SNMP) to troubleshoot problem areas and determine fault locations. These functions worked well with small, relatively simple networks, but as companies added more devices to their networks, they needed a way to collect all that management data at a central location so they could better determine what was happening on the network as a whole, rather than on individual segments.

Central network management systems, such as Hewlett-Packard's OpenView and IBM's NetView, emerged to fill this void. Meanwhile, networks continued to grow larger and more complex, so companies looked for ways to lower their management costs. Many found they were supporting two groups of network technicians, one supervising the performance of network devices and a second supervising the system's components. Rather than continue with two systems and hire people with different skill sets, *information technology* (IT) organizations wanted to consolidate them into one group. So, desktop features, such as asset management, software distribution, license monitoring, and desktop management, became priorities.

Now organizations are adding wireless devices to extend their wired networks without having to install cable. This presents yet another requirement for network management systems.

Management in the Wired World

In the wired world, network management begins with such basic hardware components as data sets (*channel service units* [CSUs]/*data service units* [DSUs]), routers, switches, multiplexers, and servers (see Figure 9-1). These and other devices typically have the ability to monitor performance, test for, and diagnose problems regarding their own operation and are capable of reporting their status to a central management station, either periodically or on request. The management station operator can initiate test procedures on systems at any point in the network. On more complex multipoint and multidrop configurations, the capability to test and diagnose problems from a central location greatly facilitates problem resolution. This capability also minimizes the need to dispatch technicians to remote locations and reduces maintenance costs.

A minimal network management system consists of a *central processing unit* (CPU), system controller, operating system software, storage device, and operator's console. The central processor may consist of a minicomputer

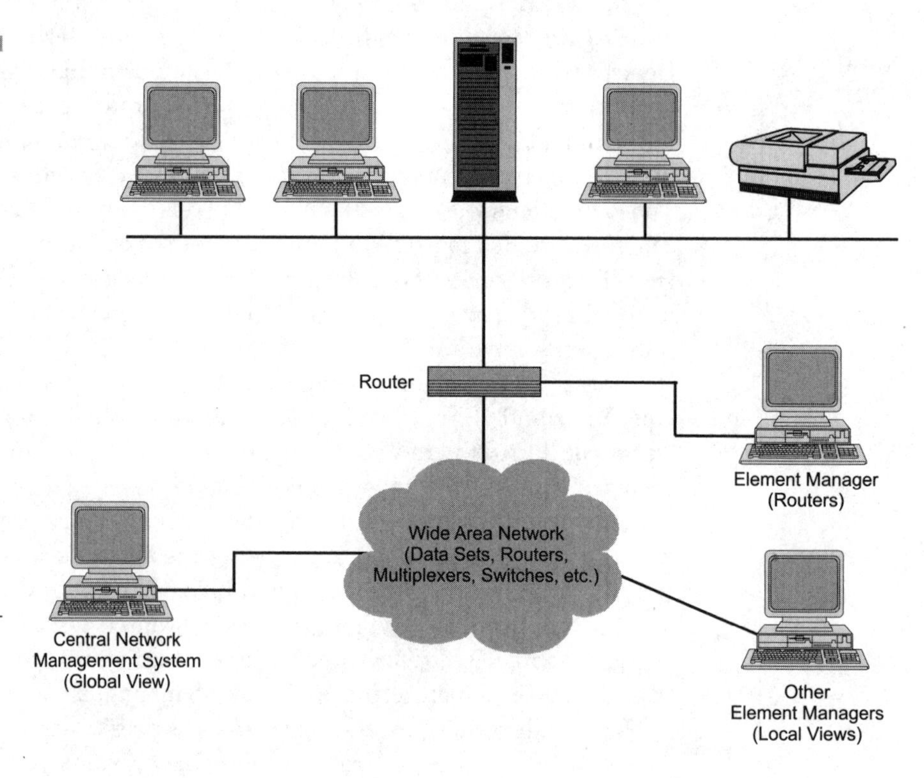

Figure 9-1 Each type of device on the network may have its own Element Management System (EMS), which reports to a Central Network Management System (CNMS) that integrates, prioritizes, and permits the analysis of information from multiple element managers.

or microcomputer. The system controller, the heart of the network management system, continuously monitors the network and generates status reports from data received through various network components. The system controller also isolates network faults and restores segments of the network that have failed or are in the process of degrading. The controller usually runs on a powerful platform such as UNIX or Windows NT/2000.

SNMP is the basis for most network management systems. It provides the ability to manage network devices in a multivendor environment from a central location. In fact, virtually any enterprise-class device that can be connected to a network supports SNMP, including wireless client adapters, *access points* (APs), and bridges. This means SNMP can be used to manage both the wired and wireless elements of an enterprise network. SNMP has several main components, including a management station, agents, and *management information bases* (MIBs).

Management Station

The manager is a program that may run on one or more hosts, with each responsible for a particular subnet. The management station provides the user interface, in the form of a *command-line interface* (CLI) or a *graphical user interface* (GUI), which provides the means to configure, monitor, analyze, and control the various components on the network.

SNMP communicates network management data to a single site called a *management station*. Under SNMP, each network segment must have a component, called an *agent*, which can monitor devices (called *objects*) on that segment and report the information to the management station. The agent may be a passive monitoring device whose sole purpose is to read the network, or it may be an active device that performs other functions as well, such as bridging, routing, and switching. Devices that are non-SNMP compliant must be linked to the management system via a proxy agent.

The manager provides the information display, communication with agents, information filtering, and control capabilities. The agents and their appropriate information are displayed in a graphical format, often against a network map. Network technicians and administrators can query the agents and read the responses on the management console. The manager also periodically polls the agents, searching for anomalies. Detection of an anomaly results in an alarm at the management system.

Management Agents

SNMP provides get, set, trap, and other functions to, among other things, retrieve, set device values, and receive notifications of network events. An agent that resides on a network device responds to requests from the management station and generates events (traps) based on instructions issued from the management station or programmed into the agent. To have a wireless AP send SNMP traps, for example, the network administrator enters the *Internet Protocol* (IP) address of the target AP along with the trap that defines the event or condition to be detected by the AP's SNMP agent. The AP might send back a notification when any of the following events occur:

- AP is powered on (cold-start trap).
- An Ethernet network connection is established (network link-up trap).
- A user has tried to communicate with the AP using an incorrect SNMP community string (authentication trap).

Another type of agent is the proxy agent, which is a program used to support devices that do not have an SNMP implementation available. The proxy is an SNMP management agent that services requests from the management station on behalf of one or more non-SNMP devices.

Management Information Base (MIB)

An MIB is a compact database for a given network component that contains the parameters that can be managed for a network device. A standard definition of an MIB exists for every device that is supported by SNMP, and a standard extension has been established called MIB II. Also, a mobile MIB is available that includes wireless-related (wireless performance statistics) and other information such as the type of network adapter and version of firmware. Via the agent, the management station monitors and updates the values in the MIB.

The MIB is a list of objects necessary to manage devices on the network. As noted, an object refers to hardware, software, or a logical association such as a connection or virtual circuit. The attributes of an object might include the number of packets sent, routing table entries, and protocol-specific variables for IP routing. A basic object of any MIB is sysDescr, which is a textual description of the entity. This value includes the full name and version identification of the system's hardware type, software operating system, and networking software. This object should contain only printable ASCII characters.

The first MIB was primarily concerned with IP routing variables used for interconnecting different networks. The core of the standard SNMP MIB is made up of 110 objects. The latest generation of MIB, known as MIB II, defines over 160 objects. It extends SNMP capabilities to a variety of media and network devices, marking a shift from Ethernets and *Transmission Control Protocol* (TCP)/IP *wide area networks* (WANs) to all media types used on *local area networks* (LANs) and WANs. Many vendors want to add value to their products by making them more manageable, so they create private extensions to the standard MIB, which can include 200 or more additional objects.

Many vendors of SNMP-compliant products include MIB toolkits that generally include two types of utilities. The first type, an MIB compiler, acts as a translator that converts ASCII text files of MIBs for use by an SNMP management station. The second type of MIB tool converts the translator's output into a format that can be used by the management station's applications or graphics. These output handlers, also known as MIB editors or MIB walkers (see Figure 9-2), let users view the MIB and select the variables to be included in the management system. Some vendors of SNMP management stations do not offer MIB toolkits, but rather an optional service whereby they will integrate into the management system any MIB a user requires for a given network. This service includes debugging and technical support.

Also, certain MIB browsers enable network managers, technicians, and systems engineers to query a remote device for software and hardware configurations via SNMP and make changes to the remote device. The remote device could be a router, switch hub, server, firewall, or any other device that supports SNMP. Another common use for an MIB browser is to find out which MIBs and *object identifiers* (OIDs) are supported on a particular device, per Figure 9-2.

Remote Network Monitor

The common platform from which to monitor multivendor networks is SNMP's *Remote Monitoring* (RMON) MIB. Although a variety of SNMP MIBs collect performance statistics to provide a snapshot of events, RMON

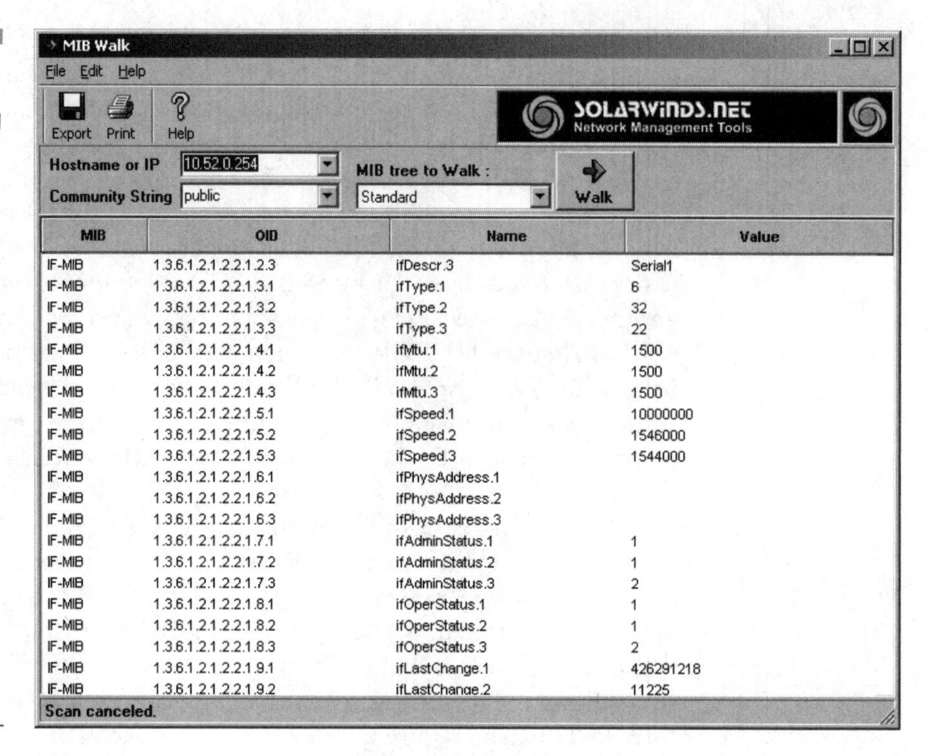

MIB	OID	Name	Value
IF-MIB	1.3.6.1.2.1.2.2.1.2.3	ifDescr.3	Serial1
IF-MIB	1.3.6.1.2.1.2.2.1.3.1	ifType.1	6
IF-MIB	1.3.6.1.2.1.2.2.1.3.2	ifType.2	32
IF-MIB	1.3.6.1.2.1.2.2.1.3.3	ifType.3	22
IF-MIB	1.3.6.1.2.1.2.2.1.4.1	ifMtu.1	1500
IF-MIB	1.3.6.1.2.1.2.2.1.4.2	ifMtu.2	1500
IF-MIB	1.3.6.1.2.1.2.2.1.4.3	ifMtu.3	1500
IF-MIB	1.3.6.1.2.1.2.2.1.5.1	ifSpeed.1	10000000
IF-MIB	1.3.6.1.2.1.2.2.1.5.2	ifSpeed.2	1546000
IF-MIB	1.3.6.1.2.1.2.2.1.5.3	ifSpeed.3	1544000
IF-MIB	1.3.6.1.2.1.2.2.1.6.1	ifPhysAddress.1	
IF-MIB	1.3.6.1.2.1.2.2.1.6.2	ifPhysAddress.2	
IF-MIB	1.3.6.1.2.1.2.2.1.6.3	ifPhysAddress.3	
IF-MIB	1.3.6.1.2.1.2.2.1.7.1	ifAdminStatus.1	1
IF-MIB	1.3.6.1.2.1.2.2.1.7.2	ifAdminStatus.2	1
IF-MIB	1.3.6.1.2.1.2.2.1.7.3	ifAdminStatus.3	2
IF-MIB	1.3.6.1.2.1.2.2.1.8.1	ifOperStatus.1	1
IF-MIB	1.3.6.1.2.1.2.2.1.8.2	ifOperStatus.2	1
IF-MIB	1.3.6.1.2.1.2.2.1.8.3	ifOperStatus.3	2
IF-MIB	1.3.6.1.2.1.2.2.1.9.1	ifLastChange.1	426291218
IF-MIB	1.3.6.1.2.1.2.2.1.9.2	ifLastChange.2	11225

Figure 9-2 MIB Walk from SolarWinds is used to search the SNMP tree for a target device and dump the value of each OID. This tool is commonly used to find out which MIBs and OIDs are supported on a particular device. MIB Walk uses a database of MIB information to determine the plain language name for each OID and the MIB to which it belongs.

enhances this monitoring capability by keeping a past record of events that can be used for fault diagnosis, performance tuning, and network planning. The original RMON MIB defines a framework for the remote monitoring of Ethernet. Subsequent RMON MIBs extend this framework to token ring and other types of networks.[1]

An RMON probe is installed on at least one remote system for each remote LAN or segment to be monitored, whether it is wired or wireless (802.11a and 802.11b). The probe views every packet and produces summary information on various types of packets, such as undersized packets, and events, such as packet collisions. The probes can also capture packets according to predefined criteria set by the network manager or test technician. At any time, the RMON probe can be queried for this information by

[1]This book is only concerned only with Ethernet because *Wireless Fidelity* (WiFi) is based on Ethernet technology. Ethernet has overtaken all other LAN technologies because it is both inexpensive and scalable to much higher speeds.

a network management application or an SNMP-based management console so that detailed analysis can be performed in an effort to pinpoint where and why an error occurred.

Despite the interoperability of WiFi equipment from different vendors, when it comes to monitoring *wireless LANs* (WLANs), vendors of RMON probes/consoles support only certain types of equipment. Some vendors only support WiFi equipment from Cisco, Intel, and Nortel. The reason is that the WiFi vendors use different manufacturers' chipsets, and these chipsets must support RMON. To complicate matters, a different chipset may be used for 802.11a and 802.11b products. In short, before wireless products can be monitored with RMON, they must be designed to support it. This further differentiates wireless products aimed at consumers and those intended for the enterprise.

Using RMON

A management application that views the internetwork, for example, gathers data from RMON agents running on each segment in the network, both wired and wireless. The data is integrated and correlated to provide various internetwork views that provide end-to-end visibility of network traffic, both LAN and WAN. The operator can switch between views to monitor the performance of the network from a variety of perspectives.

For example, the operator can switch among a *Media Access Control* (MAC) view (which shows traffic going through routers and gateways), a network view (which shows end-to-end traffic), or a view in which, by applying filters, only the traffic of a given protocol or a suite of protocols is shown. These traffic matrices provide the information necessary to configure or partition the internetwork to optimize LAN and WAN utilization.

In selecting the MAC-level view, for example, the network map shows each node of a segment separately, indicating intrasegment node-to-node data traffic. It also shows the total intersegment data traffic from routers and gateways. This combination enables the operator to see consolidated internetwork traffic and how each end node contributes to it.

In selecting the network-level view, the network map shows end-to-end data traffic between nodes across segments. By connecting the source and ultimate destination, without clouding the view with routers and gateways, the operator can immediately identify specific areas contributing to an unbalanced traffic load.

Another type of application enables the network manager to consolidate and present multiple segment information, configure RMON alarms, and perform baseline measurements and long-term reporting. Alarms can be set on any RMON variable, and notification via traps can be sent to multiple management stations. Baseline statistics enable long-term trend analysis of network traffic patterns that can be used to plan for network growth.

Ethernet Object Groups

The original RMON specification consists of nine Ethernet groups, which are discussed later. The drawback to this implementation of RMON is that it is capable of collecting information up to the MAC layer only. A follow-up specification called RMON-II enables probes to gather information from the network layer (Layer 3) all the way up to the application layer (Layer 7) of the *Open Systems Interconnection* (OSI) reference model. The ramifications of this are that RMON-II probes can gather not only the information its predecessor can, but also information about network-specific applications like the *Hypertext Transfer Protocol* (HTTP), *Network File System* (NFS), AppleTalk, and *Internetwork Packet Exchange* (IPX). With this added data collection capability, administrators can obtain a more complete picture of the enterprise network.

Ethernet Statistics Group The Statistics Group provides segment-level statistics (see Figure 9-3). These statistics show packets, octets (or bytes), broadcasts, multicasts, and collisions on the local segment, as well as the number of occurrences of dropped packets by the agent. Each statistic is maintained in its own 32-bit cumulative counter. Real-time packet size distribution is also provided.

Ethernet History Group With the exception of packet-size distribution, which is provided only on a real-time basis, the History Group provides historical views of the statistics provided in the Statistics Group. The History Group can respond to user-defined sampling intervals and bucket counters, allowing for some customization in trend analysis.

The RMON MIB comes with two defaults for trend analysis. The first provides for 50 buckets (or samples) of 30-second sampling intervals over a period of 25 minutes. The second provides for 50 buckets of 30-minute sampling intervals over a period of 25 hours. Users can modify either of these or add additional intervals to meet specific requirements for historical analysis. The sampling interval can range from one second to one hour.

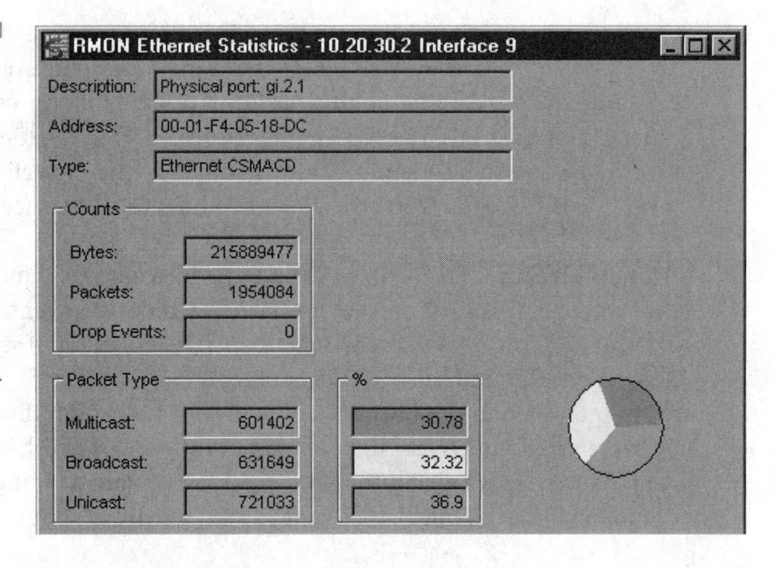

Figure 9-3 The Ethernet Statistics window accessed from Enterasys Networks' NetSight Element Manager. This window can be used to view a detailed statistical breakdown of traffic on the monitored Ethernet network segment. The data provided applies only to the interface or network segment, either wired or wireless.

Host Table Group The RMON MIB specifies a host table that includes node traffic statistics: packets sent and received, octets sent and received, and broadcasts, multicasts, and errored packets sent. In the host table, the classification "errors sent" is the combination of undersized packets, fragments, *Cyclic Redundancy Check* (CRC)/alignment errors, collisions, and oversized packets sent by each node.

The RMON MIB also includes a host timetable that shows the relative order in which the agent discovered each host. This feature is not only useful for network management purposes, but it also assists in uploading those nodes to the management station of which it is not yet aware. This reduces unnecessary SNMP traffic on the network.

Host Top N Group The Host Top N Group extends the host table by providing sorted host statistics, such as the top 10 nodes sending packets or an ordered list of all nodes according to the errors sent over the last 24 hours. The data selected and the duration of the study are both defined at the network management station. The number of studies that can be run depends on the resources of the monitoring device.

When a set of statistics is selected for study, only the selected statistics are maintained in the Host Top N counters; other statistics over the same time intervals are not available for later study. This processing, performed remotely in the RMON MIB agent, reduces SNMP traffic on the network and the processing load on the management station, which would otherwise need to use SNMP to retrieve the entire host table for local processing.

Alarms Group The Alarms Group provides a general mechanism for setting thresholds and sampling intervals to generate events on any counter or integer maintained by the agent, such as segment statistics, node traffic statistics defined in the host table, or any user-defined packet match counter defined in the Filters Group. Both rising and falling thresholds can be set, each of which can indicate network faults. Thresholds can be established for both the absolute value of a statistic or its delta value, enabling the manager to be notified of rapid spikes or drops in a monitored value.

Filters Group The Filters Group provides a generic filtering engine that implements all packet capture functions and events. The packet capture buffer is filled with only those packets that match the user-specified filtering criteria. Filtering conditions can be combined using the Boolean parameters "and" or "not." Multiple filters are combined with the Boolean "or" parameter.

Packet Capture Group The type of packets collected is dependent upon the Filter Group. The Packet Capture Group enables the user to create multiple capture buffers and to control whether the trace buffers will wrap (overwrite) when full or stop capturing. The user may expand or contract the size of the buffer to fit immediate needs for packet capturing, rather than permanently committing memory that will not always be needed.

Notifications (Events) Group In a distributed management environment, the RMON MIB agent can deliver traps to multiple management stations that share a single community name destination specified for the trap. In addition to the three traps already mentioned—rising threshold, falling threshold (refer to "Alarms Group"), and packet match (refer to "Packet Capture Group")—seven additional traps can be specified:

- **coldStart** This trap indicates that the sending protocol entity is reinitializing itself such that the agent's configuration or the protocol entity implementation may be altered.
- **warmStart** This trap indicates that the sending protocol entity is reinitializing itself such that neither the agent configuration nor the protocol entity implementation is altered.

- **linkDown** This trap indicates that the sending protocol entity recognizes a failure in one of the communication links represented in the agent's configuration.
- **linkUp** This trap indicates that the sending protocol entity recognizes that one of the communication links represented in the agent's configuration has come up.
- **authenticationFailure** This trap indicates that the sending protocol entity is the addressee of a protocol message that is not properly authenticated. Although implementations of the SNMP must be capable of generating this trap, they must also be capable of suppressing the emission of such traps via an implementation-specific mechanism.
- **egpNeighborLoss** This trap indicates that an *External Gateway Protocol* (EGP) neighbor for whom the sending protocol entity was an EGP peer has been marked down and the peer relationship is no longer valid.
- **enterpriseSpecific** This trap indicates that the sending protocol entity recognizes that some enterprise-specific event has occurred.

The Notifications (Events) Group enables users to specify the number of events that can be sent to the monitor log. From the log, any specified event can be sent to the management station. The log includes the time of day for each event and a description of the event written by the vendor of the monitor. The log overwrites when full, so events may be lost if not uploaded to the management station periodically.

Traffic Matrix Group The RMON MIB includes a traffic matrix at the MAC layer. A traffic matrix shows the amount of traffic and the number of errors between pairs of nodes: one source and one destination address per pair. For each pair, the RMON MIB maintains counters for the number of packets, octets, and error packets between the nodes. Users can sort this information by source or destination address.

Applying remote monitoring and statistics-gathering capabilities to the Ethernet environment offers a number of benefits. The availability of critical networks is maximized, since remote capabilities enable a more timely resolution of the problem. With the capability to resolve problems remotely, operations staff can avoid costly travel to troubleshoot problems on site. By being able to analyze data collected at specific intervals over a long period of time, intermittent problems can be tracked down that would normally go undetected and unresolved. With RMON II, these capabilities are enhanced and extended up to the applications level across the enterprise.

RMON II

As noted, the RMON MIB is basically a MAC-level standard. Its visibility does not extend to the router port, meaning that it cannot see beyond individual LAN segments. As such, it does not provide visibility into conversations across the network or connectivity between the various network segments. Given the trends toward remote access and distributed workgroups that rely on wireless as well as wired connections, both of which may generate a lot of intersegment traffic, visibility across the enterprise is an important capability to have.

RMON II extends the packet-capture and decoding capabilities of the original RMON MIB to Layers 3 through 7 of the OSI reference model. This allows traffic to be monitored via network-layer addresses, letting RMON see beyond the router to the internetwork, and distinguish between applications based on the protocols used. Since a wireless bridge does not perform any IP routing or forwarding, however, certain groups of managed objects are not meaningful. For SNMP requests pertaining to such managed objects, the node simply returns a "no such name" error status in the response.

Analysis tools that support the network layer can sort traffic by protocol rather than just reporting on aggregate traffic. This means that network managers will be able to determine, for example, the percent of IP versus IPX traffic traversing the network. In addition, these higher-level monitoring tools can map end-to-end traffic, giving network managers the ability to trace communications between two hosts, or nodes, even if the two are located on different LAN segments. The following RMON II functions enable this level of visibility:

- The *protocol directory table* provides a list of all the different protocols an RMON II probe can interpret.
- The *protocol distribution table* permits tracking on any given segment the number of bytes and packets that have been sent from each of the protocols supported. This information is useful for displaying traffic types by percentage in graphical form.
- *Address mapping* permits identifying traffic-generating nodes, or hosts, by Ethernet or token ring addresses in addition to MAC address. It also discovers switch or hub ports to which the hosts are attached. This is helpful in node discovery and network topology applications for pinpointing the specific paths of network traffic.
- The *network-layer host table* permits tracking bytes, packets, and errors by host according to individual network-layer protocols.

- The *network-layer matrix table* permits tracking the number of packets sent between pairs of hosts by network-layer address.
- The *application-layer host table* permits tracking bytes, packets, and errors by host according to the application.
- The *application-layer matrix table* permits tracking conversations between pairs of hosts by application.
- The *history group* permits filtering and storing statistics according to user-defined parameters and time intervals.
- The *configuration group* defines standard configuration parameters for probes that include such parameters as network address, serial line information, and SNMP trap destination information.

RMON II is focused more on helping network managers understand traffic flow for the purpose of capacity planning rather than for the purpose of physical troubleshooting. However, the capability to identify traffic levels and generate statistics by application has the potential to greatly reduce the time it takes to troubleshoot certain problems. Without tools that can pinpoint which software application is responsible for gobbling up a disproportionate share of the available bandwidth, network managers can only guess. Often it is easier just to upgrade a server or a buy more bandwidth, which inflates operating costs and shrinks IT budgets.

GUI SNMP Tools

Being a simple application, SNMP does not have all the features of a full-blown network management system. Among the many capabilities SNMP does not have are automatic discovery and topology mapping. The automatic discovery capability finds and identifies all devices or nodes connected to the network. Based on the discovered information, the network management system automatically populates a topology map. Nodes that cannot be discovered automatically can be represented by manually adding custom or standard icons to the appropriate map views, or by using the network management systems' SNMP-based APIs for building map applications without having to manually modify the configuration to accommodate non-SNMP devices.

A network map is useful for ascertaining the relationships of various equipment and connections, keeping an accurate inventory of network components, and isolating problems on the network. The network map is updated automatically when any device is added or removed from the network. A device status is displayed via color changes to the map. Any changes to the network map are carried through to the relevant lower-layer maps.

Some vendors have developed graphical wrappers for SNMP, which makes SNMP easier to use than its traditional text-based CLI. They have also appended additional functionality such as automatic discovery and topology mapping (see Figure 9-4), giving SNMP-based products the look and feel of a high-end network management system.

Many wireless devices, including both APs and bridges, support SNMP. This is a standard protocol for managing a variety of systems and network devices on both wired and wireless networks. A set of common management tasks can be done with SNMP, but vendors often include proprietary management features in their products that can also be handled via SNMP. This enables SNMP to perform some management tasks that cannot be done using the vendor's GUI. For example, it might be possible to disable SSID broadcasts from an AP using SNMP but not with the vendor's GUI configuration tool.

But care must be taken when using SNMP. Products often ship with a default password. When changing other default settings, administrators should not overlook changing the default SNMP password (often called the

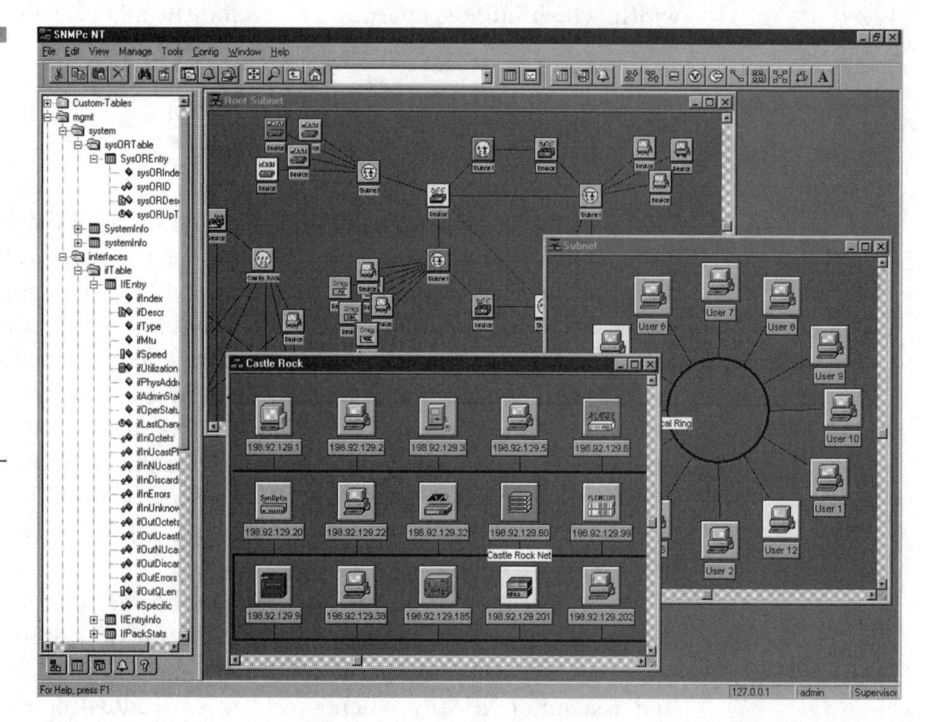

Figure 9-4 Castle Rock Computing, Inc. offers SNMPc for Windows NT and Windows 2000, which, among other things, provides a graphical display that supports multilevel hierarchical mapping.

community) as well. As an extra security precaution, administrators should consider disabling SNMP until it is actually needed. If a hacker ever succeeds in exploiting the security vulnerabilities of SNMP, he or she could tie up network resources at will (discussed later).

The Mechanics of SNMP

SNMP specifies a structure for formatting messages and for transmitting information between reporting devices and data-collection programs on the network. The SNMP-compliant devices on the network are polled for performance-related information, which is passed to a network management console. Alarms are also passed to the console, where the gathered information can be viewed to pinpoint problems on the network or be stored for later analysis.

SNMP runs on top of TCP/IP's datagram protocol—specifically, the *User Datagram Protocol* (UDP), a transport protocol that offers a connectionless-mode service. This means that a session need not be established before network management information can be passed to the central control point. Although SNMP messages can be exchanged across any protocol, UDP is well suited to the brief request/response message exchanges characteristic of network management communications.

As noted, SNMP is a flexible network management protocol that can be used to manage virtually any object. As stated earlier, an object refers to hardware, software, or a logical association, such as a connection or virtual circuit. An object's definition is written by the equipment vendor and is held in an MIB. The MIB is simply a list of switch settings, hardware counters, in-memory variables, or files used by the network management system to determine the alarm and reporting characteristics of each device on the network, including those connected over LANs.

Also noted previously is that SNMP is basically a request/response protocol. The management system retrieves information from the agents through SNMP's *get* and *get-next* commands. The get request retrieves the values of specific objects from the MIB, which lists the network objects for which an agent can return values. These values may include the number of input packets, the number of input errors, and routing information. The get-next request permits navigation of the MIB, enabling the next MIB object to be retrieved, relative to its current position. A set request is used to request a logically remote agent to alter the values of variables. In addition to these message types, *trap* messages can occur, which are unsolicited messages conveyed from a management agent to management stations. Other commands are available that enable the network manager to take specific actions to control the network. Although these commands look like SNMP commands, they are really vendor-specific implementations. For example, some vendors use a *stat* command to determine the status of network connections.

All the major network management platforms support SNMP. In addition, many of the third-party systems and network management applications that plug into these platforms also support SNMP. The advantage of using such products is that they take advantage of SNMP's capabilities while providing a GUI to make SNMP easier to use. Even MIBs can be selected for display and navigation through the GUI.

Another advantage of commercial products is that they can use SNMP to provide additional functionality. For example, Hewlett-Packard's OpenView is used to manage network devices that are IP addressable and run SNMP. The automatic discovery capability finds and identifies all IP nodes on the network, including those of other vendors that support SNMP. On the basis of discovered information, the management system automatically draws a network topology map. Nodes that cannot be discovered automatically can be represented in either of two ways: first, by manually adding custom or standard icons to the appropriate map views, and second, by using SNMP-based APIs for building map applications without having to manually modify the configuration to accommodate non-SNMP devices.

SNMP on Wireless Networks

Although they share standard elements and mechanisms, wired and wireless networks have significant differences. In addition to the conventional wired network, wireless networks have the following unique issues:

- **Secondary hierarchy** The wireless network environment is hierarchical, with portable devices associated to a given AP. Standard network management products designed for wired networks often do not represent this critically important tiered topology network structure and are more likely to represent the wireless network as a flat topology.
- **Roaming** The wireless network environment supports dynamic cell connection or roaming, which is the process of changing the network connection of a portable device from one AP to another as the user changes location.

- **Persistence of mobile units** In an environment comprised of portable devices, the persistence of the devices also becomes a factor. Unlike desktop systems or other network components that operate continuously or are powered on and off daily, portable devices typically are turned on and off frequently throughout the day, making it difficult to monitor them.
- **SNMP agents** The capability of devices on wired and wireless networks to host an SNMP agent also impacts network management functionality. Desktop or notebook computers have adequate amounts of memory and processing power to support an SNMP agent operating as a background task handling requests from the management station. With *personal digital assistants* (PDAs) and other handheld devices, however, memory and processing power are limited resources, making it difficult to provide agents for these devices.

The Performance of SNMP

Managing a wireless network is a significantly more difficult task than managing a wired network. One reason has to do with the unpredictable behavior of the wireless channel due to fading, jamming, and atmospheric conditions. This variability of signal quality can reduce the efficiency of the management operation. The bandwidth of wireless links is another issue. Bandwidth will always be limited due to the properties of the physical medium and regulatory limits on the use of radio spectrum, making it necessary for network protocols to utilize the available bandwidth efficiently.

Using mobile agents to facilitate management offers several benefits, including reduced network traffic and the efficient utilization of processing resources. On the other hand, the use of mobile agents consumes considerable resources from the agent host, which can pose a problem for mobile terminals like PDAs, and the response time in notifying the network manager depends on the wireless link quality. The response time can vary from several seconds to even several minutes depending on the link condition. Management functionality is unusable in situations with bad link quality and with system load at 20 percent of the whole capacity. Unusable means not being able to update the whole MIB tree using SNMP's get-next operation. Accordingly, a management station can manage several times the number of devices when the signal strength is good than when it is bad with the same system load. The response time increases in an exponential way as the load of the system grows.

Using SNMP's *get-bulk* request, on the other hand, provides the means to retrieve much more data about a network entity—as much data as is possible to carry in the response message. This is especially useful when channel conditions are bad because the number of round-trip timeouts is reduced, which is important in maintaining performance in a wireless environment. With traffic reduced between manager and agent, the probability of UDP packet loss is also much less.

Although measurements obtained through the get-next command are good for short transactions, such as retrieving individual objects when the query and response messages fit into a single UDP packet, if the MIB contains a considerable amount of tables, the retrieval process will involve a high number of *protocol data unit* (PDU) exchanges over the network. In the case of bad link quality, this could make management performance worse, but when using a single get operation, it is much easier to adapt channel conditions for the current transaction. The user can specify the relevant options such as timeout for the response message or the maximum number of retransmissions with respect to link quality.

The SNMP agent itself can provide an approximate indication of the link condition. For instance, information about the number of retransmissions performed by the manager/agent application or the maximum delay time for receiving a response measured by the manager could be useful for determining WLAN performance.

Platform Support for Wireless

IT departments tend to favor a single management console that integrates wireless management with existing wired infrastructure. Yet with so much change in the wireless industry and the rapid pace of standards development, many established network management platform vendors do not yet support wireless networks. Instead, they rely on wireless vendors or third parties to provide management applications that can be incorporated into their management platform to manage the wireless component of the network. With its simplicity, SNMP provides management capabilities from a central point of control that spans wired and wireless environments.

Network management systems have been specifically designed for the wireless environment, enabling administrators to navigate, examine, and manage the wireless network Such wireless management systems may even offer seamless integration into larger corporate enterprise network management platforms, such as Hewlett-Packard's OpenView Network Node Manager, to provide a single management environment. This integration offers IT managers the tools to install, configure, centrally monitor, and manage the entire enterprise, wired and wireless. This approach also helps IT departments avoid the additional expense of purchasing, training, and maintaining a new management tool.

Basic Wireless Management Features

A good wireless network management system should offer the following basic features:

- Automated discovery of new APs and remote configuration capabilities.
- AP configuration that includes a group management capability, a configuration database, and standard SNMP support.
- Configuration of the *access control list* (ACL) to regulate which devices are allowed to associate with an AP.
- Threshold and event notification whereby designated events trigger alerts before a system problem or failure occurs. There should also be real-time updates on network events.
- Exception-based management that enables key parameters in the WLAN to be monitored for values that exceed a predefined threshold, so network operators or support personnel can be notified.
- Single-source report generation for inventory, configuration, and usage.
- Data trending to track and trend network traffic and usage, and proactive analysis of network utilization.
- Mapping to provide a concise network representation, including a tree view showing all devices in various levels of detail. There should also be a facility for creating custom maps and importing device-specific images.
- Drill-down and component-zooming capabilities, enabling information to be expanded as required.
- Intelligent agent technology that monitors and coordinates tracking information from multiple data sources, thereby reducing the amount of management traffic on the network.
- Remote firmware and software revision control to permit the configuration of system components from a central location without having to tie up staff resources with onsite service calls.
- A wireless proxy agent to support non-SNMP mobile devices such as PDAs.

SNMP Security Issues

As useful as SNMP is for managing systems and network devices on wired and wireless networks, it has always had weak security, making it vulnerable to hackers. In particular, the decoding and subsequent processing of management messages between SNMP managers and agents may cause *denial of service* (DoS) conditions, service interruptions, and buffer overflows, allowing an attacker to gain access to or disrupt the proper operation of a device affected by these vulnerabilities. To create a DoS attack, for example, a hacker might send bogus SNMP requests and traps that could flood an SNMP management system or appliance running a trap application. This might cause the system to hang and may require a reboot.

Usually, vendors offer SNMP security patches or firmware updates free of charge, and they can be downloaded from their web sites. If the SNMP-managed equipment was purchased from a vendor that went out of business, however, a fix might not be available. Rather than risk an attack, it is better to replace the equipment with a newer model from a stable vendor. In addition to applying a patch from the SNMP vendor to prevent these problems from reoccurring, network administrators can take the following additional precautions:

- Enable ingress filtering, which is the blocking of access to SNMP services at the network perimeter.
- Configure SNMP agent systems to disallow request messages from nonauthorized systems.
- Segregate SNMP traffic onto a separate management network and filter traffic as it leaves the network to prevent the network from being used as a source for attacks on other sites.
- Disable any service or capability that is not explicitly required, including SNMP if it is enabled.

The last point is very important because SNMP will help malicious users learn a lot about a target system, making password guessing and similar attacks much easier. Unless this service is required, it is highly recommended that SNMP be turned off. Since many vendors ship their products with SNMP enabled, this security hole can only be plugged by using the configuration utility to disable SNMP.

If SNMP is implemented and it is not responding, taking the following steps may resolve the issue:

- Check the community name. Community names are case sensitive and it is general practice to use uppercase letters for community names.
- If all the community names are removed, including the default name (Public), SNMP will not respond to any community names presented.
- If the existing SNMP settings are changed, they take effect immediately. However, if SNMP is being configured for the first time, the SNMP service must be restarted before the settings can take effect.
- Check to see if the host is set to accept SNMP packets only from specific sources.
- Make sure that the Send Authentication Trap option is enabled, and verify the list of trap destinations.
- Check the permission level of both community rights and the MIB access variable. The most restricted setting is used.

Conclusion

SNMP's popularity stems from the fact that it works and is reliable, widely supported, and extendible. The protocol itself is now in the public domain, and SNMP capabilities have been integrated into just about every conceivable device that is used on today's LANs and WANs, whether they are wired to the network or linked wirelessly to the network. Some vendors offer graphical management systems for their wireless products. Cisco's Wireless Manager is one example. It is a UNIX-based solution that can be run locally from a server console in a UNIX workstation or a Windows computer or remotely via *remote login* (rlogin), Telnet, or *remote shell* (rsh).

Wireless devices can even be managed through a web browser. Although the web management interface may not support the remote configuration of any security-sensitive parameters, its value is in giving IT staff more flexibility to monitor network performance from any location that has Internet access. To implement web management, the browser proxy is configured for a direct Internet connection. The administrator then enters the IP address of the target AP in the browser window. The administrator is prompted for a username and password. Once connected, a navigation tree appears on the left side, providing access to the management features.

CHAPTER 10

Selecting a Wireless Internet Service Provider

Companies interested in providing their traveling employees with wireless services so they can access e-mail, the Internet, and resources on the corporate network face a bewildering array of choices among service providers. This type of provider is referred to as a *wireless Internet service provider* (WISP), which allows subscribers to connect to a server using medium-range wireless links. This type of ISP offers broadband service and allows subscribers to access the Internet and the Web from anywhere within the zone of coverage provided by the server antenna. This is usually a region with a radius of several miles.

Different types of WISPs are entering the market, each experimenting to find a viable business model. Since the market for wireless services is still a nascent one, the WISP business is highly fragmented to the point where users must subscribe to multiple services to get wireless connectivity in the locations they travel to most.

Market Landscape

WISPs come from a variety of backgrounds. Some are ISPs that have implemented wireless technology to extend the reach of their services in markets underserved by *Digital Subscriber Lines* (DSLs) or cable. Some are mobile operators that are looking to combine 2.5/3G wireless technologies with *Wireless Fidelity* (WiFi) to give their customers more connectivity options based on where they happen to be and what applications they may be using at any given time. Telephone companies and interexchange carriers are also looking to get into the wireless market by deploying WiFi technologies. In addition, infrastructure operators, city networks, enterprises, property owners, and managed communities have set up wireless networks for themselves and may or may not be charging for the service. Many entrepreneurs have entered this market as well, capitalizing on the low cost of entry. Because of their different backgrounds, the business models of these players will vary. The market for 802.11 wireless services is so new that a clearly successful business model has yet to emerge.

It is uncertain how the market will respond to questions such as how different WISPs and owners of hot spot locations will coexist and cooperate, especially with regard to roaming. Many experts believe that roaming is the key to the success of the WISP market. Although roaming agreements between WISPs will likely contribute to market success in the long run, spotty coverage and lack of roaming may be obstacles for many users in the short term. It will take time to build a national wireless infrastructure capable of providing meaningful coverage and connecting all the hot spots.

Service pricing in the emerging wireless market is also an issue. Long-term service contracts, month-to-month service, and even daily rates have been established to accommodate users who do not travel often, but who need Internet access when they do go on the road. Some wireless providers offer products and services for free at zero margin or subzero margin. These strategies are usually aimed at buying market share, possibly with the idea of transitioning to a profit model after users become addicted and find they can no longer do without their wireless connections. This is fine in theory, but this strategy also begs the question, "How many consumers would continue to pay for a service they had been getting for free?"

Although some idealists out there posit a "free Internet" model and see wireless technology as a means of realizing this goal, the cost of equipment, maintenance, power, and eventually a wired connection to the Internet backbone at a *network access point* (NAP) all conspire against this ideal, making it unworkable. An initial and ongoing cost is associated with any type of service infrastructure, which no amount of idealism can eliminate for very long. Business users demand stability in their service providers. They do not choose free services because the future viability of the arrangement is always in doubt and it is not so easy to start all over again with another provider.

Related to the issue of seamless mobility is the problem of billing for roaming services and revenue sharing between operators. Although customer billing and revenue sharing are clear-cut among telephone companies and interexchange carriers for conventional voice and data services, much work needs to be done in these areas for wireless services. As demonstrated by *voice over Internet* (VoIP) services between multiple ISPs, however, this is not an insurmountable obstacle for WISPs. Although it did take about five years to set up and smooth out billing and revenue sharing among ISPs selling voice services, the delay was due in large part to the initial resistance of telephone companies to VoIP. With 802.11 wireless services, however, these issues probably will not take as long to hash out because the telephone companies and national mobile operators are very much interested in offering WiFi services themselves, so much can be gained by devising billing systems and revenue-sharing formulas.

Types of Providers

Several types of WISPs exist: aggregators, hot spot operators, facility owners, wide area WISPs, and network community WISPs. Each is more or less suited to the emerging wireless connectivity needs of businesses. These categories are fluid and one service provider may span several categories.

Aggregators

Aggregators offer technical solutions required by other operators to provide wireless Internet services to their customers. They provide equipment and systems, wireless access, interconnection facilities, security, roaming capabilities, and billing and settlement services. They may also offer network integration services for the connection of different wireless networks, including those based on such differing technologies as *General Packet Radio Service* (GPRS) and WiFi. Businesses typically do not buy products and services directly from aggregators, but more aggregators are actively pursuing enterprise-class customers who travel frequently.

Aggregators or wholesalers construct WiFi networks, offering shared bandwidth to service providers and other customers. In the case of service providers, the use of a shared infrastructure allows them to focus on marketing, user administration, and billing functions. The easiest way for the providers to bill users for their wireless services is by credit card, which minimizes the costly administrative staff required to send out invoices and follow up on delinquent accounts. The aggregator pays for fixed broadband access and may also set up the wireless APs in its markets. The result is that anyone can become an operator of a wireless network and offer wireless Internet services to the public under their own brand and pricing scheme without actually having to invest in infrastructure.

iPass and GRIC Communications are examples of aggregators. Both companies started in the aggregation business by providing ISPs the means to offer Internet roaming services to their dial-up customers. This arrangement allows member ISPs to offer customers local dial-up access to the Internet from around the world using their standard ID and password. Later, as VoIP became popular, iPass and GRIC extended this arrangement to providers of IP telephony service. Now they have extended the concept further to include 802.11b wireless services. As the number of hot spots and areas covered by different wireless networks increases, so does the importance of this kind of cooperation and collaboration for implementing a roaming capability.

The problem for independent WISPs is that they cannot afford to build a global or even national service footprint on their own. Aggregators fill this void. The aggregator not only fills in the coverage gaps of smaller ISPs and offers a range of access methods to their customers, but it also delivers applications such as e-mail, conferencing, messaging, expense management, and other specialized enterprise applications via a universal client that is loaded on the end user's computer. The interface provides a directory of services and APs around the world.

In the case of GRIC, reliable and redundant coverage of 20,000-plus dial-up and wireless local APs is provided in over 150 countries with its GRIC MobileOffice solution. This is the user's universal wireless and dial-up interface to the GRIC network, which makes it easy to access wireless hot spots in airports, hotels, convention centers, and other facilities. When a user is near a WiFi roaming hot spot, the location can be selected from the hot spot list accessible from the GRIC client interface (see Figure 10-1). In places where a wireless hot spot is not available, the user can gain Internet access through a dial-up AP listed in the GRIC client interface.

iPass offers a comparable service and client interface called iPassConnect (see Figure 10-2). The aggregator has redundant APs and redundant network backbones in major business centers throughout the world supporting dial-up, *Integrated Services Digital Network* (ISDN), *Personal Handyphone System* (PHS), broadband, and wireless services. The iPass phonebook contains over 14,000 unique APs in 150 countries.

Not only does the aggregator manage the shared network, but it also provides a clearinghouse for revenue settlements. Clearinghouse functions are based on common technical, service, and payment standards to settle charges that the independent operator's customers incur when they access the network facilities of another operator to initiate Internet roaming, for example. Using software and services provided by the aggregator, operators are able to manage and settle Internet roaming transactions, identify and authorize end users to conduct those transactions, and determine preferred

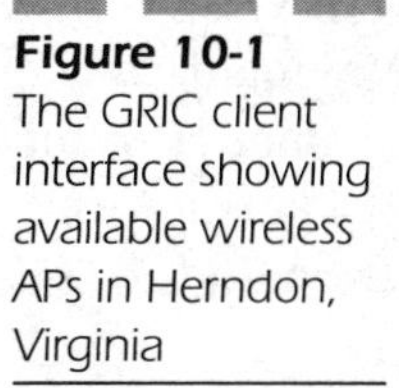

Figure 10-1 The GRIC client interface showing available wireless APs in Herndon, Virginia

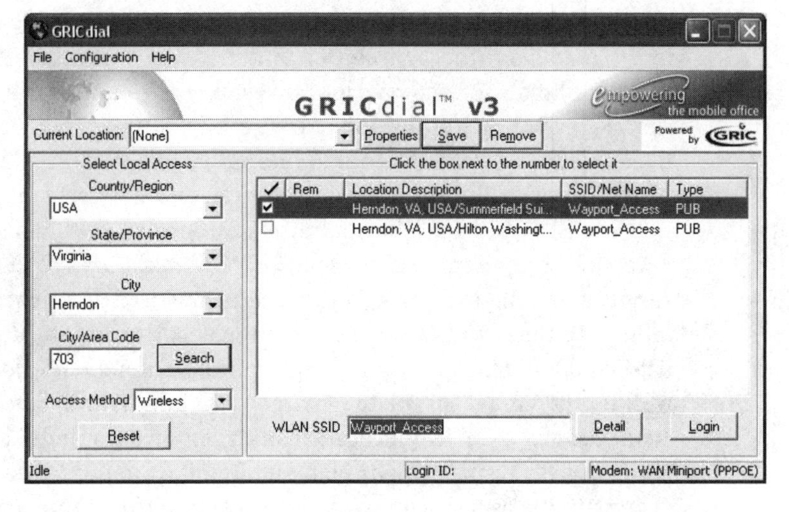

Figure 10-2 The iPassConnect client interface showing the methods of access available to roaming users: modem, ISDN, PHS, wireless broadband (that is, 802.11) and wired broadband

routing for Internet-based communications. For each global Internet roaming transaction, the aggregator tracks usage, collects the amount that a roamer's home service provider owes, pays the appropriate amount to the service provider enabling local access, and provides underlying usage data to the home operator to enable billing of its end user. Ultimately, the end user receives a single bill for all services and networks used.

With GPRS and WiFi capabilities integrated onto a single PC Card for notebook computers and *personal digital assistants* (PDAs)—and as both services become available in more locations—authentication, security, and billing systems will evolve to encompass both services. This means that a user could go from accessing data on a GPRS network, for example, while walking down the street to getting that data over a much faster wireless connection in a public place, such as a restaurant or hotel lobby. Charges for both services would appear on a single bill from the home service provider, even though the services are actually coming from other sources through an aggregator.

Businesses can subscribe to the services of independent operators at relatively low risk if the operator has a relationship with an aggregator. Risk is minimized because the operator will have had to pass the aggregator's rigorous quality assurance and operations testing regimen. The operator will also have to be certified as enterprise ready in terms of such key factors as interoperability with the leading *virtual private networks* (VPNs), personal firewalls, and other enterprise-mandated security systems.

In terms of security, one of the hardest tasks facing enterprise *information technology* (IT) managers is securing computers when they are not in the office. Aggregators address this issue with personal firewall software that is loaded to the computers of traveling employees. The aggregator also monitors the personal firewall while the user is remotely connected. If the personal firewall is ever shut down, the user will be immediately disconnected from both the VPN as well as the aggregator's network. These measures provide IT managers with assurance that their traveling or telecommuting employees will be securely connecting to their corporate *local area networks* (LANs) even while roaming from the home network.

From the perspective of the independent service provider, a relationship with an aggregator can be leveraged in several ways to make its services more attractive to enterprise customers:

- The ability of the aggregator to bundle services allows independent operators to cross-sell other Internet-based services to end users and reduce end-user turnover.
- The relationship with the aggregator allows independent operators to expand their geographic reach. By being able to access an existing network, independent operators avoid the cost and time required to negotiate, build, and maintain their own network of relationships.
- Through an aggregator, the small operator can provide businesses with Internet access using any commercially viable technology, including WiFi. As the aggregator's portfolio of wired and wireless Internet access and other technologies becomes available, the small operator can take advantage of them to offer additional services to business customers.
- Independent operators can better compete for enterprise customers by expanding their service offerings without incurring the capital investments and other costs necessary to build and deploy their own global Internet-based communications network.
- The aggregator offers redundant points of presence. By offering multiple Internet connection alternatives in most locations, it enables business customers of the smaller operator to connect regardless of most single points of network failure.

- An aggregator provides a high degree of security using industry-standard protocols for all Internet connections made over its network. The aggregator's client software implements the security features, including VPNs. This enables the smaller operator to provide secure services to businesses without having to employ security specialists.
- Affiliating with an aggregator gives the smaller operator access to all the networks of other operators. The aggregator's technical, service, and payment standards facilitate network sharing by enhancing business and technical interoperability.

To ensure the quality of its network, the aggregator's network management center monitors the performance of each Internet AP. A suite of tools enables the real-time monitoring of network events and it facilitates the management of configurations, problems, performance, security, and reporting. This information enables the aggregator to identify the highest-quality Internet APs, update routes, and remove poor-quality Internet APs from its electronic phonebook. It also permits the generation of custom trending and *quality of service* (QoS) reports for its network member operators.

Hot Spot WISPs

Another type of WISP is the hot spot operator. Here the focus is on providing wireless coverage in public spaces such as airports, hotels, convention centers, restaurants, and cafés where the target group can get Internet access by paying for it on an as-needed basis or by subscription. This strategy entails installing radio base stations within physical facilities so that users can access the Internet wirelessly when inside the building. These WISPs have a relationship with the facility owner that lets them install equipment and operate within its premises. They typically strike a deal with a landlord to deploy wireless APs in the facility and then pay the landlord monthly fees and/or a cut of the revenue.

Businesses might consider service subscriptions to hot spot operators for their traveling employees. Many companies have arrangements with hotels for a corporate rate, and hot spot operators have installed a wireless capability in these hotels. Wayport is an example of a hot spot operator. It serves almost 500 major hotels, including Four Seasons, Loews, Wyndham, Hilton, Marriott, Sheraton, Doubletree, and Embassy Suites.

Of course, all these hotels accommodate dial-up Internet access, providing a convenient hookup of a notebook computer's modem to a telephone's built-in data port. Through a local ISP number or a corporate 800 number, employees can get the usual 56 Kbps service. But wireless broadband service offers business travelers data throughput at rates comparable to what they are accustomed to in their office. Consequently, businesses are reassessing their hotel policies based on the availability of broadband wireless Internet access, both to eliminate unnecessary phone charges and help traveling employees stay productive, especially when transferring large files. The selection of service provider will be based primarily on what hotels the company already has relationships with.

Instead of using wireless Internet service as a profit center, most facility owners view it as a key differentiator that results in a short-term competitive advantage. Business travelers are more apt to stay at a hotel, especially for extended periods, if it has advanced services that permit them to maintain a high degree of communications, collaboration, and productivity while they are away from the office. In the long run, wireless Internet access will become a standard service just like a TV in each room.

The same applies to airports. Many companies have travel policies that stipulate the use of preferred airlines. Employees benefit by these policies because it gives them the opportunity to accumulate bonus miles, qualifying them for free tickets when certain thresholds are met. Usually, companies allow employees to keep these bonus miles for their personal use. Many times frequent flyers will also sign up for club membership, which entitles them to use the airline's private lounge where they are offered premium accommodations while waiting for flights. Increasingly, these accommodations include wireless Internet access as a further inducement to join.

Wayport provides wireless service at such airports as Dallas-Fort Worth, Seattle-Tacoma, San Jose, Austin-Bergstrom, and Oakland. At these airports, travelers can also take advantage of Wayport's business centers, which feature private offices with high-speed Internet access and other business services, as well as mobile devices and accessories for sale.

Boingo Wireless is an example of a hot spot aggregator. Through its partnerships with multiple wireless service providers, Boingo offers service in hundreds of major business-class hotels, including Four Seasons, Hilton, Marriott, Sheraton, Radisson, and Wyndham hotels. Boingo also offers select coverage at airports such as Atlanta-Hartsfield (ATL), Chicago-Midway (MDW), Philadelphia (PHL), Baltimore-Washington (BWI), Washington-Dulles (IAD), and Boston-Logan (BOS).

Instead of deploying its own network infrastructure, Boingo purchases networks from 802.11b operators on a wholesale basis, integrates these networks together, and sells a single service to its customers. The advantage of this model to business customers is that it saves them from having to purchase multiple subscriptions. If a user at Los Angeles Airport (LAX) walks from the American Airlines Admiral's Club to the Delta Crown Room, for example, he or she will not need to sign up with two different wireless service providers. By signing up with Boingo, the user can take advantage of wireless connectivity at both locations. Boingo takes care of the revenue split between the different providers.

By bringing together these fragmented networks, Boingo is able to drive more traffic to their networks and increase demand for their services. Boingo provides the marketing, technical support, end-user software, and billing, while the hot spot operators build and operate the WiFi networks. Boingo even makes it easy for entrepreneurs to quickly set up their own hot spot locations under their own brand name, allowing them to offer customized hot spot services to property owners. They can create your own initial sign-in screen with their own logo, offer free-trial coupons and prepaid cards, and determine their own end-user pricing. In addition, because the solution is Boingo ready, the entrepreneur would be able to offer its customers access to other Boingo partner hot spots and be paid by Boingo for such network traffic. As the Boingo service expands in this way, business customers benefit by having to subscribe to only one service, regardless of their travel destination.

Facility Owner WISPs

Although a building can be a hot spot for wireless service, a facility owner WISP implies much more. This is a property owner that offers wireless services to business tenants for a profit. Although the facility owner may outsource wireless Internet access to a service provider, many operate as WISPs, especially when communications services are a core business activity. In such cases, the facility owner may want to provide customers with an attractive service package where wireless Internet access is merely one of many components. For example, an office building owner may already be offering tenants long distance, private lines, DSL, ISDN, and a variety of other services, including local calling through a shared *private branch exchange* (PBX) as an added value to office space. These and other services are purchased at wholesale rates and sold at competitive retail rates, which are better than those that could be obtained by the tenants individually if they had to seek their own sources of equipment and services.

The building owner already has technical staff to perform moves, adds, and changes to keep up with tenants' needs. Adding wireless service to the mix of services already offered is not beyond their skill set. In fact, the building owner views wireless as an opportunity to make money from current tenants and as a means of attracting new tenants. The facility owner does, however, depend on the existence of hardware vendors, network integrators, and consultants to provide elements of the overall infrastructure. In fact, the facility owner masters the technical competencies involved in setting up and managing the wireless infrastructure, bearing all installation, upgrade, and expansion costs. The building owner can then partner with a hot spot aggregator like Boingo Wireless to offer subscribers a roaming capability to their travel destinations, providing even more value to current and prospective tenants.

Facility owners can deliver wireless service incrementally to serve only tenants that want to subscribe to it, or they can deploy a wireless solution throughout the building to provide complete wireless broadband access at a cost that is less than that of wiring the building. By setting up a subnet for each tenant, maximum data throughput can be provided and security can be enforced. This also provides flexibility for meeting the needs of demanding administrators who want to build temporary networks to support conferences or workgroups. With the entire building covered with a wireless broadband network, even visitors can connect wirelessly anywhere in the building. Companies looking for office space may want to add the availability of wireless service to their selection criteria.

The business tenant benefits from this arrangement by having access to advanced services and receiving one itemized invoice for office space, communications services, conference room usage, audio/visual equipment, coffee service, and administrative services. This model seems to work better when the tenants are smaller companies; larger firms may opt out of the building owner's shared service offerings, if they have their own technical staff and value having control of their own network. Companies might want control over their own communication systems out of concern for security, to retain the ability to mix and match best-of-breed products, to closely monitor network performance, to ensure timely responses to impending system or network failures, and to select their own staff based on critical needs. When added together, this kind of control translates into a competitive advantage over companies that are dependent upon others for the proper operation and management of their communication systems and networks.

Wide Area WISPs

As the term implies, wide area WISPs provide wireless coverage over a larger geographical area than hot spots. They deliver broadband access to homes and offices, providing coverage inside and outside of buildings, as well as in public places such as parks and recreation areas. Technically, these are hot spots, and the wide area WISP can act as an aggregator to interconnect these hot spots to create a larger network through which users can roam. Although companies like Boingo Wireless, GRIC Communications, and iPass have been discussed from the perspective of aggregators, they could also be considered examples of wide area WISPs.

To date, the concept of the wide area WISP has yet to be fully realized. This may change soon, however, as large companies start to pool resources to build a nationwide high-speed wireless data network. In late 2002, AT&T, IBM, and Intel formed Cometa Networks with the intent of selling WiFi service on a national scale. Throughout 2003, the company will work with major national and regional retail chains, hotels, universities, and real-estate firms to deploy the broadband wireless access service in hot spots in the top-50 U.S. metropolitan areas.

Cometa Networks plans to provide this service to telecommunications companies, ISPs, cable operators, and wireless carriers, who can then offer their customers wireless Internet access. Cometa will also offer wireless Internet access to enterprise customers through the participating carriers.

Cometa's service will make it possible for users to keep their existing sign-on procedures, e-mail addresses, IDs, passwords, and payment methods—regardless of whether they are accessing the Internet via an ISP, corporate VPN, telecommunications provider, or cable operator.

AT&T will provide the network infrastructure and management. IBM will provide wireless site installations and back-office systems. Intel's strategic investment program, Intel Capital, will provide most of the investment funds to finance the new network.

Although Boingo Wireless and others are already building out national networks by linking hot spots, they do not come close to having the resources of Cometa. The participation of large companies, particularly cellular carriers, is essential to the success of any nationwide WiFi network, since they have the customer base to which such services could be readily marketed. The feasibility of a nationwide WiFi network is made possible by the rapid buildup of the critical mass needed to support such a network, with many notebooks and PDAs featuring built-in WiFi chips and antennas. The WiFi Alliance estimates that shipments of WiFi cards and access clients combined are running at 1.5 million units per month.

As it is now, the public hot spot market is fragmented, with WLANs interspersed around cities and in coffee shops, airport lounges, and hotel lobbies. Weaving a truly national wireless network is beyond the capabilities of any single company. The participation of such large companies in Cometa Networks brings considerable resources to such a venture. In addition to investment capital, the partnership brings extensive network expertise to create a viable business model for a cohesive, nationwide wireless network, providing customers with easy roaming as they move from one hot spot to another.

In general, cellular carriers have ample incentive to throw their weight behind WiFi. Although the number of cellular subscribers is still growing, it is slowing as the market reaches saturation. Faced with stagnating growth and the continued delay of 3G adoption, wireless carriers have an incentive to seriously explore WiFi. They must find a way to drive up revenue in the future, and WiFi is one of the few technologies with a fast growth curve. Unlike 3G services, which required a huge investment in licenses, WiFi operates in unlicensed frequency bands, eliminating that cost. Another fact about WiFi that cellular carriers find compelling is that a clear market demand exists for such services, whereas the market demand for 3G services is still in doubt.

As Cometa's national WiFi network becomes available, it would be worthwhile for businesses to consider using it. Instead of being limited to a building or hot spot, or the scattered coverage provided by an aggregator, a true roaming capability would ensure wireless connectivity over the wide area and, possibly, a handoff of services between WiFi and 3G.

Network Community WISPs

Among WISPs is a group that employs a totally different strategy to provide wireless Internet services: network community WISPs. These operators are dedicated to providing free service. Although this category of WISP may provide hot spot coverage or wide area coverage, they are distinguished from hot spot operators and wide area WISPs by their open-source orientation, which stipulates that the Internet should be free and accessible to everyone.

One of the largest network community WISPs is SeattleWireless, which believes that setting up a community-based wireless network should be as easy and inexpensive as possible. The stated goals of SeattleWireless are, "To quantitatively foster community growth and cohesion through wireless networks; to enable and support community organizations (emergency, academic, non-profit, and social), while allowing for improved localized and remote communications; and to build networks that are self-sustaining." The guiding philosophy of SeattleWireless is to have the network owned by the people who create it, and not have to report or be responsible to entities that would like to fund it.

These network communities do not focus on a specific target group, as many other WISPs tend to do. Certainly, they do not focus on the business segment, considering the lack of security within the network communities. The business segment often has higher demands for security, whereas the wireless Internet is insecure by default. These communities also do not plan to charge for services in the future, as this goes against the open-source concept. In terms of marketing, such networks gain users by word of mouth and expand as necessary; aside from a web page, no specific marketing activities take place. Funding comes from donations and from the voluntary efforts of people who want to feel part of a larger scheme of things. This enables the network community WISP to remain independent and self-reliant.

Another large network community WISP is NYCwireless, which works with businesses, government agencies, and nonprofit organizations to help develop free wireless Internet access throughout the New York metropolitan area, including Bryant Park. Thanks to a series of unobtrusive antennas scattered throughout the park, anyone with a wireless-enabled notebook computer can log onto the network for an unlimited time. Bryant Park is in the middle of 300 million square feet of office space that constitutes the midtown Manhattan office market. Thousands of people can take advantage of the wireless network by occasionally working outside as if they were in their own backyard.

Because of the uncertain viability of not-for-profit organizations, corporations would not usually rely on this type of wireless operator for running business applications. Since independent volunteers run each AP with their own equipment, settings may vary from node to node. Users must refer to the specific web page of that node operator and check its coverage map. The default settings are an *Extended Service Set Identifier* (ESSID)/Network name www.nycwireless.net, and the *Dynamic Host Configuration Protocol* (DHCP) is used to obtain an IP address.

The openness of such networks poses security concerns for business users. Even NYCwireless advises users to implement the *Secure Sockets Layer* (SSL) protocol to connect to web pages and mail hosts, to use *Secure Shell* (SSH) instead of Telnet whenever possible, and to implement VPNs for all other data to ensure privacy and security. The danger for business users is that they may not be familiar with security issues, and by using this type of network, they may inadvertently leave the door open for entry to the corporate network by hackers.

Periodically, debate surfaces about the noncharge issue. Within the network community, ultimately someone is paying for the bandwidth that gets wireless traffic to the Internet. The provider of an AP may already have the device for other purposes, such as home, business, or school use, so the additional cost is absorbed elsewhere. For example, a cable or DSL subscriber may use WiFi in the home for networking various computers and providing them with shared access to the Internet through the cable or DSL service. The AP collects the traffic from the computers in the home and routes it to the cable of the DSL connection. Since the AP radiates its signal out into the street, the subscriber can let anyone use it to gain free access to the Internet through that cable or DSL service. In effect, the cable or DSL service provider's network is being used by people who are not its subscribers.

The practice of redistributing bandwidth opens moral and legal issues. Most service agreements for DSL or cable modems prohibit the user from carrying other people's traffic, although this kind of bandwidth sharing is difficult for ISPs to detect. Many network community WISPs try to avoid these problems by recommending that AP contributors check their service agreements for specific prohibitions against bandwidth sharing. They also have an acceptable use policy that all users must agree with, partly to protect the people who open their Internet access to wireless users from liability for any illegal activities in which those users might engage.

Some national ISPs, such as AT&T and Time Warner, specifically prohibit bandwidth sharing with a WiFi network. In New York, Time Warner Cable sent out letters to users who publicly identified their networks as free WiFi hot spots on a local web site. The letter demanded that users immediately discontinue offering open access and warned that using the account in that manner violates the subscription agreement. The letter also mentioned a $50,000 civil remedy for noncompliance.

Not all ISPs agree with this stance. Covad, for example, supports community WiFi networks and does not object if a user has more bandwidth than he or she needs and wants to share it. Boingo also supports free community networks. If an operator chooses not to charge for access, Boingo will promote the AP in its client software location database and will not charge its customers to access the community network.

In summary, network community WISPs are not oriented toward meeting enterprise needs. In fact, these WISPs pride themselves on not being beholden to corporate interests. Their focus is on finding ways to offer free Internet access, rather than on providing security and other special features. Given these considerations, plus the potential legal liabilities, it is recommended that companies revise their security policies to warn employees about using their corporate notebooks to access such networks. These networks are inherently insecure and could be used by hackers to gain entry into the corporate network.

Wireless Application Service Providers (WASPs)

The advantages of providing a mobile workforce with seamless, anytime, anywhere access to corporate data and applications are obvious. Less clear is how smaller companies can deploy wireless data systems in a timely, economical manner. A large enterprise has virtually limitless opportunities it can pursue, ranging from creating custom applications in house to hiring an outside software firm to design custom applications to run on a wireless network. But smaller firms that cannot afford to develop custom wireless solutions can turn to a *wireless application service provider* (WASP).

For the most part, WASPs rent rather than sell applications. This allows companies to access their main systems—e-mail, financials, manufacturing, *customer relationship management* (CRM), *enterprise resource planning* (ERP), and other enterprise applications—wirelessly. WASPs host the rented or leased applications offsite, which is done through their own secure servers. This allows business customers to save money on internal IT staffing and other resources.

For example, with a hosted wireless solution, companies do not have to hire wireless-savvy IT staffers or train their existing IT staff on wireless technologies. They can be up and running more quickly than when creating and installing an in-house wireless solution. And as a service, hosting means companies do not need to worry about what protocols or devices to use for running wireless applications; the hosting provider will handle these and other security and network technology issues. The hosting provider is better equipped to upgrade to new technologies, add new devices and functionality, and scale up applications to meet growing user needs. WASPs also offer predictable costs plus guaranteed service quality.

Among the largest WASPs is AppShop, the world's largest Oracle application service provider. AppShop's services include the Mobile Oracle E-Business Suite, which enables wireless subscribers on various carrier networks to securely access their enterprise Oracle data. Using a wireless switch, Appshop's wireless capability grants customers secure access to their Oracle applications through a variety of wireless devices, including PDAs, two-way messaging devices, one-way pagers, and *Wireless Application Protocol* (WAP) mobile phones.

Oracle's e-business applications, such as CRM, e-commerce, ERP, finance, human resources, manufacturing, marketing, product development, supply chain management, and transportation, are easily accessible to users in the field and away from the office through their existing mobile devices.

The wireless offering not only offers the standard *graphical user interface* (GUI) through which access to the applications is granted over wireless links, but it can also be enhanced with voice and audio capabilities for easier data interaction while on the move. The system ensures security by using 128-bit encryption technology, and users access the application via a unique username and password, affording multilevel authentication. AppShop's basic data access wireless solutions cost in the range of $50 to $100 per user, per month.

It is important to select the right type of WASP for a given application. WASPs fall into several categories:

- WASPs that simply repurpose existing web pages for wireless devices. This type of WASP can be useful when all the organization wants to do is augment its web site with wireless capabilities to allow mobile users to access content.
- Wireless content providers that collect and/or build and deliver web-based content to wireless devices, such as news, stock updates, travel schedules, sports scores, and other time-sensitive information.
- Wireless infrastructure providers that provide hosting, middleware, and network gateway services among networks and devices.
- Traditional application service providers that host enterprise applications, but that also offer wireless access to those applications.

Depending on the number and type of applications an organization has, it may enter into relationships with several WASPs to meet all of its needs. Before examining WASP offerings, it is important to perform an internal needs survey. If a given WASP formats content for WAP phones but no other devices, for example, this could limit the success of the company's wireless project. Although millions of phones in use are WAP enabled, different formats might be in use throughout the organization. A certain amount of users will demand WAP, but others might use a palm device, a WinCE-based PDA, or a Research In Motion Blackberry pager, for example.

In addition to the types of mobile devices used, the geographical distribution of employees should be considered as well because 2.5/3G network protocols differ by region. In North America, smart phones with PDA capabilities may be used on networks based on either *Code Division Multiple Access* (CDMA) or *Time Division Multiple Access* (TDMA), whereas in

Europe, the *Global System for Mobile Communications* (GSM) is used. In Asia, the WASP may need to support i-Mode. Although these mobile devices may also support WiFi with the addition of a card, the user may have to resort to 2.5/3G services when WiFi service is not available. The more ways employees can make wireless connections, the more valuable that network is to the people who use it.

Businesses contemplating arrangements with WASPs, however, should be mindful of the potential trade-offs. Although WASPs offer several advantages, they have numerous disadvantages as well. First among the disadvantages is the loss of control that comes with outsourcing applications. Larger customers tend to prefer in-house solutions because it allows them to make rapid changes to their applications and to leverage their existing IT investments. Companies are also concerned about the security issues involved in taking mission-critical applications outside the enterprise, especially when access to those applications occurs over wireless connections.

The success of the relationship largely depends on choosing the right partner. Many companies run into trouble by picking the wrong WASP and find themselves in a compromising position. The way to sidestep many of these issues is by checking out potential service providers ahead of time and going through the basic due diligence process.

In addition to basic queries about pricing, security processes, strategic partners, customer references, and the financial stability of the WASP, companies also need to find out what happens if things do go wrong. Is the application being outsourced such that it can be reclaimed for support by the in-house staff? Is there a *service level agreement* (SLA) that describes minimum system uptime and penalties for substandard performance? Does the WASP provide a clear upgrade path for the application so the company's employees have access to all the latest features and functions? If the answers are too vague or blatantly satisfactory, it is best to look elsewhere.

WASPs present a viable option for taking a company wireless. If they offer easy-to-deploy, low-cost solutions and have a track record of success, they may be worthy of serious evaluation. Despite all the marketing hype about wireless services and applications, the industry faces some serious issues that may hinder ubiquitous service, including

- **Fragmentation among wireless service providers** Services today are offered primarily by small, privately held WISPs that are focused on building market share by aggressively installing equipment one location at a time.

- **Undercapitalization** Many of these firms require continuous infusions of investment capital to continue building out their networks. Without adequate funding, customer service may be lacking at a critical time.
- **Unproven business models** It is still not clear how business partners in WiFi services can make money. After equipping several hundred Starbucks locations and other properties with wireless connectivity, MobileStar, for example, filed for bankruptcy and sold its assets to VoiceStream.
- **Lack of roaming agreements** Currently, service providers only offer islands of wireless connectivity. Even with roaming agreements and the involvement of aggregators, coverage is still spotty and is likely to remain so for the foreseeable future.

On the other hand, roaming is an advantage that DSL and cable providers cannot currently offer. The competitive advantage of offering a business customer seamless, high-speed roaming, even to limited venues, may make all the difference to businesses whose employees require more than a dial-up connection.

Conclusion

WISPs offer customized high-speed Internet service packages for businesses. The high cost of other business-level Internet options often forces businesses to make due with a shared dial-up Internet connection. The high-powered connectivity, continuous online access, networking capabilities, and the affordability of WISP services, however, make them ideal for enterprise applications. Businesses can use WISP services to send and receive files quickly, research information, connect with customers, place orders online, and improve productivity.

WASPs combine the traditional services of application service providers and WISPs. WASPs, however, do more than simply host wireless applications for enterprise customers. Many WASPs provide a software platform that enables them to offer a middleware service, whereby a company's back-end legacy system can be linked over multiple kinds of wireless networks. They will then deliver content from those systems in multiple formats to a variety of handset models. For example, among the many middleware services a WASP will offer is to WAP-enable an application. Sometimes all the customer has to do from a technical standpoint is provide the WASP with an API to the back-end application it wants to mobilize, which can often be obtained from the software vendor. In this and other ways, WASPs do a good job of shielding companies from the development complexities of linking applications to the myriad types of wireless networks and enabling the application to run across a range of mobile device types.

ACRONYMS

ACL Access control list

ACL Asynchronous connectionless (link)

ADPCM Adaptive Differential Pulse Code Modulation

ADSL Asymmetrical Digital Subscriber Line

AES Advanced Encryption Algorithm

AGC Automatic gain control

AM Amplitude modulation

AMPS Advanced Mobile Phone Service

ANSI American National Standards Institute

AP Access point

API Application programming interface

ARP Address Resolution Protocol

ARPA Advanced Research Projects Agency

AS Autonomous system

ASCII American Standard Code for Information Interchange

ASP Application service provider

ATIS Alliance for Telecommunications Industry Solutions (formerly ECSA)

ATM Asynchronous Transfer Mode

AWG American wire gauge

BER Bit error rate

BERT Bit Error Rate Tester

BIOS Basic input/output system

BOC Bell Operating Company

BootP Boot Protocol

BPS Bits per second

CAN Campus area network

CCA Clear channel assessment

CD-ROM Compact disc read-only memory

CDMA Code Division Multiple Access

CDPD Cellular Digital Packet Data

CERT Computer Emergency Response Team

CLEC Competitive Local Exchange Carrier

CLI Command-line interface

CO Central office

CoA Care of address

CoS Class of service

CPE Customer premises equipment

CPS Cycles per second (Hertz)

CPU Central processing unit

CRC Cyclic Redundancy Check

CRM Customer relationship management

CS Cell station

CS-CDPD Circuit-Switched Cellular Digital Packet Data

CSC Circuit-switched cellular

CSCCP Circuit-Switched CDPD Control Protocol

CSD Circuit-switched data

CSMA/CA Carrier sense multiple access with collision avoidance

CSMA/CD Carrier sense multiple access with collision detection

CTS Clear to Send

CVSD Continuously variable slope delta (modulation)

D-AMPS Digital Advanced Mobile Phone Service

DA Destination address

dB Decibel

DBMS Database Management System

DBS Direct Broadcast Satellite

DC Direct current

DCC Digital control channel

DCCH Digital control channel

DCE Data communications equipment

DECT Digital Enhanced (formerly European) Cordless Telecommunication

DES Data Encryption Standard

DHCP Dynamic Host Configuration Protocol

DIP Dual inline pin

DLL Data link layer

DMA Direct memory access

DMI Desktop management interface

DMTF Desktop Management Task Force

DMZ Demilitarized zone

DNS Domain Name Service

DoD Department of Defense (U.S.)

DOS Disk Operating System

DS0 Digital signal level 0 (64 Kbps)

DS1 Digital signal level 1 (1.544 Mbps)

DS1C Digital signal level 1C (3.152 Mbps)

DS2 Digital signal level 2 (6.312 Mbps)

DS3 Digital signal level 3 (44.736 Mbps)

DS4 Digital signal level 4 (274.176 Mbps)

DSL Digital Subscriber Line

DSP Digital signal processor

DSSS Direct sequence spread spectrum

DTE Data terminal equipment

E-AMPS Enhanced Advanced Mobile Phone Service

E-mail Electronic mail

E-TDMA Expanded Time Division Multiple Access

EAP Extensible Authentication Protocol (Cisco Systems)

ECSA Exchange Carriers Standards Association

ED Ending delimiter

ED Energy-detect

EDGE Enhanced Data Rates for Global Evolution

EHF Extremely high frequency (more than 30 GHz)

EIA Electronic Industries Association

EIR Equipment identity register

EISA Extended Industry Standard Architecture

EMI Electromechanical interference

EP Extension point

ERP Effective radiated power
ERP Enterprise resource planning
ESD Electronic software distribution
ESMR Enhanced Specialized Mobile Radio
ESMS Enhanced Short Message Service
ESMTP Extended Simple Mail Transfer Protocol
ESN Electronic serial number
ESSD Extended Service Set Identifier
ETSI European Telecommunication Standards Institute
EVRC Enhanced variable rate coder
FA Foreign agent
FACCH Fast associated control channel
FC Frame control
FCC Federal Communications Commission (U.S.)
FCS Frame check sequence
FDD Frequency division duplexing
FDDI Fiber Distributed Data Interface
FHSS Frequency-hopping spread spectrum
FIR Fast Infrared
FM Frequency modulation
FOCC Forward control channel
FS Frame status
FT1 Fractional T1
FTP File Transfer Protocol
FWA Fixed wireless access
FWT Fixed wireless terminal
GEO Geostationary earth orbit
GHz Gigahertz (billions of cycles per second)
GMT Greenwich Mean Time
GPRS General Packet Radio Service
GPS Global Positioning System
GSM Global System for Mobile Telecommunications (formerly Groupe Spéciale Mobile)

GUI Graphical user interface

HA Home agent FA

HDML Handheld Device Markup Language

HDSL High-bit-rate Digital Subscriber Line

HEC Header error check

HF High frequency (1.8 MHz to 30 MHz)

HLR Home location register

HomeRF Home Radio Frequency

HSCSD High-Speed Circuit-Switched Data

HTML HyperText Markup Language

HTTP HyperText Transfer Protocol

Hz Hertz (cycles per second)

I/O Input/output

IAD Integrated access device

IC Integrated circuit

ICEA Insulated Cable Engineers Association

ICF Internet Connection Firewall

ICMP Internet Control Message Protocol

ICP Integrated communications provider

ICS Internet Connection Sharing

ID Identification

iDEN Integrated Digital Enhanced Network

IEC International Electrotechnical Commission

IEEE Institute of Electrical and Electronic Engineers

IESG Internet Engineering Steering Group

IETF Internet Engineering Task Force

IF Intermediate frequency

IKE Internet Key Exchange

ILEC Incumbent Local Exchange Carrier

IMEI International mobile equipment identity

IMSI International mobile subscriber identity

IMTS Improved Mobile Telephone Service

INMS Integrated Network Management System

IP Internet Protocol

IPv4 Internet Protocol version 4 (current)

IPv6 Internet Protocol version 6 (future)

IPSec Internet Protocol Security

IPX Internet Packet Exchange

IR Infrared

IrDA Infrared Data Association

IrDA-SIR Infrared Data Association Serial Infrared (standard)

IrFM Infrared financial messaging

IrLAN Infrared local area network

IrLAP Infrared Link Access Protocol

IrLMP Infrared Link Management Protocol

IrPL Infrared physical layer

IrTTP Infrared Tint Transport Protocol

IRQ Interrupt Request

IrTTP Infrared Transport Protocol

IS Information system

IS Industry standard

IS-IS Intra-autonomous system to intra-autonomous system

ISA Industry Standard Architecture

ISM Industrial, Scientific, and Medical (frequency bands)

ISO International Organization for Standardization

ISP Internet service provider

IT Information technology

IV Initialization vector

IXC Interexchange carrier

J2ME Java 2 Micro Edition

JDBC Java Database Connectivity

JDC Japanese digital cellular

JDK Java Development Kit

JMAPI Java Management Application Programming Interface

JPEG Joint Photographic Experts Group

JVM Java Virtual Machine

K (Kilo) One thousand (Kbps)

KB Kilobyte

kHz Kilohertz (thousands of cycles per second)

L2F Layer 2 Forwarding

L2TP Layer 2 Tunneling Protocol

LAN Local area network

LCD Liquid crystal display

LD Laser diode

LDCELP Low Delay Code Excited Linear Prediction

LEAP Lightweight Extensible Authentication Protocol (Cisco Systems)

LEC Local exchange carrier

LED Light-emitting diode

LEO Low earth orbit

LF Low frequency (30 kHz to 300 kHz)

Li-Ion Lithium ion

LLC Logical Link Control

LMDS Local Multipoint Distribution System

LSI Large-scale integration

M (Mega) One million (Mbps)

M-ES Mobile end system

MA Mobility agent

MAC Media Access Control

MAC Moves, adds, changes

MAE Major economic area

MAN Metropolitan area network

MAU Multiple access unit

MB Megabyte

MD-IS Mobile Data—Intermediate System

MDBS Mobile Database System

MDLP Mobile Data-link Layer Protocol

MDS Multipoint Distribution Service

MEO Middle earth orbit

MES Master earth station

MH Mobile host

MHz Megahertz (millions of cycles per second)

MIB Management information base

MIN Mobile identification number

MM Mobility manager

MMDS Multichannel Multipoint Distribution Service

MMS Multimedia Messaging Service

MNLP Mobile Network Location Protocol

MNRP Mobile Network Registration Protocol

MO Mobile originating

Modem Modulation/demodulation

MPEG Moving Pictures Experts Group

MPLS Multiprotocol Label Switching

MR Message register

ms Millisecond (thousandths of a second)

MS Mobile station

MSC Mobile switching center

MSG Message

MSRN Mobile station roaming number

MSS Mobile satellite service

MT Mobile terminating

MTBF Mean time between failure

MTSO Mobile Telephone Switching Office

MTSO Mobile Transport Serving Office

mW Milliwatt

MXU Mobile exchange unit

N-AMPS Narrowband Advanced Mobile Phone Service

N-PCS Narrowband Personal Communications Service

NAP Network access point

NAT Network Address Translation

NAVSTAR Navigation System with Timing and Ranging

NE Network element

NEI Network equipment identifier

NEMA National Electrical Manufacturers Association
NetBIOS Network Basic Input/Output System
NFS Network File System (or Server)
NIC Network interface card
NiCd Nickel cadmium
NiMH Nickel-metal hydride
NIST National Institute of Standards and Technology
NIU Network interface unit
nm Nanometer
NM Network manager
NMS Network Management System
NMT Nordic Mobile Telephone (Ericsson)
NOC Network operations center
NOS Network operating system
NPCS Narrowband Personal Communication Services
OBEX Object exchange
OC Optical carrier
OEM Original equipment manufacturer
OFDM Orthogonal frequency division multiplexing
OMC Operations and maintenance center
OS Operating system
OSI Open Systems Interconnection
OSPF Open shortest path first
PA Preamble
PACS Personal Access Communications System
PAN Personal area network
PAP Password Authentication Protocol
PAT Port Address Translation
PBX Private branch exchange
PC Personal computer
PCB Printed circuit board
PCH Paging channel
PCM Pulse code modulation

PCMCIA Personal Computer Memory Card International Association

PCN Personal communications networks

PCS Personal communications services

PDA Personal digital assistant

PDN Packet data network

PDSN Packet data serving node

PDU Payload data unit

PGP Pretty good privacy

PHS Personal Handyphone System

PHY Physical layer

PIAF PHS Internet Access Forum (Japan)

PIM Personal information manager

PIN Personal identification number

PIN Positive-Intrinsic-Negative

PnP Plug and Play

PoE Power over Ethernet

POP Point of presence

POTS Plain Old Telephone Service

PPP Point-to-Point Protocol

PPS Packets per second

PPTP Point-to-Point Tunneling Protocol

PSN Packet-switched network

PSP Power-Saving Protocol

PSTN Public Switched Telephone Network

PT Payload type

PTT Post Telephone & Telegraph

PUC Public Utility Commission

QoS Quality of service

RACH Random access channel

RADIUS Remote Authentication Dial-in User Service

RAM Random access memory

RARP Reverse Address Resolution Protocol

RF Radio frequency

RF Routing field

RFI Radio frequency interference

RIP Routing Information Protocol

RMON Remote Monitoring

ROI Return on investment

RRM Radio resource management

RSS Received signal strength

RSSI Received Signal Strength Indicator

RT Remote terminal

RTS Request to Send

RX Receive

SA Source address

SACCH Slow associated control channel

SCO Synchronous connection-oriented

SD Secure Digital

SD Starting delimiter

SDF Service data function

SDIO Secure digital input/output

SFD Start frame delimiter

SHF Super high frequency (3 to 30 GHz)

SHTTP Secure HyperText Transfer Protocol

SID System identification

SIM Subscriber identification module

SINR Signal-to-Interference Noise Ratio

SIR Signal-to-Interference Ratio

SKU Stock Keeping Unit

SLA Service level agreement

SMS Short Message Service

SMTP Simple Mail Transfer Protocol

SNMP Simple Network Management Protocol

SNR Signal-to-noise ratio

SOHO Small office/home office

SQL Structured Query Language

SSH Secure Shell

SSID Service set identifier

SSL Secure Sockets Layer

STP Shielded twisted pair

STP Spanning Tree Protocol

T1 Transmission service at the DS1 rate of 1.544 Mbps

T3 Transmission service at the DS3 rate of 44.736 Mbps

TASI Time Assigned Speech Interpolation

TB Terabyte (trillion bytes)

Tbps Terabit per second

TCO Total cost of ownership

TCP Transmission Control Protocol

TDD Time division duplexing

TDM Time division multiplexing

TDMA Time Division Multiple Access

TDMA/TDD Time Division Multiple Access with time division duplexing

TFTP Trivial File Transfer Protocol

TKIP Temporal Key Integrity Protocol

TPDDI Twisted Pair Distributed Data Interface

TX Transmit

UDP User Datagram Protocol

UDP/IP User Datagram Protocol/Internet Protocol

UHF Ultra high frequency (238 MHz to 1.3 GHz)

UMTS Universal Mobile Telephone Service

UMTS Universal Mobile Telecommunications System

UP Unified Protocol

UPS Uninterruptible power supply

URL Uniform resource locator

USB Universal serial bus

UTP Unshielded twisted pair

UWB Ultra Wide Band

VC Virtual circuit

VF Voice frequency

VFIR Very Fast Infrared

VHF Very high frequency (50 MHz to 146 MHz)

VLAN Virtual local area network

VLF Very low frequency (less than 30 kHz)

VLR Visitor location register

VLSI Very large scale integration

VOFDM Vector orthogonal frequency division multiplexing

VoIP Voice over Internet Protocol

VPN Virtual private network

VSAT Very small aperture terminal

WAN Wide area network

WAP Wireless access point

WAP Wireless Application Protocol

WASP Wireless application service provider

WBM Web-based management

WCDMA Wideband Code Division Multiple Access

WCS Wireless Communications Service

WDCT Worldwide Digital Cordless Telephone

WECA Wireless Ethernet Compatibility Alliance (now known as WiFi Alliance)

WEP Wired Equivalent Privacy

WiFi Wireless Fidelity

WISP Wireless Internet service provider

WLAN Wireless local area network

WLL Wireless local loop

WLSE Wireless LAN Solution Engine (Cisco Systems)

WML Wireless Markup Language

WPA WiFi Protected Access

WSP Wireless Session Protocol

WWW World Wide Web

XOR Exclusive OR

INDEX

Note: *Boldface numbers indicate illustrations*

B

C

D

G

H

I

K–L

M

N

O

P

Q–R

S

T

U–V

W

Y–Z

ABOUT THE AUTHOR

Nathan Muller is cofounder and senior consultant of Ascent Solutions Group, which designs and executes innovative sales, marketing, and training support programs for technology providers and users. With 30 years of telecommunications industry experience, Mr. Muller has written extensively on many aspects of computers and communications, having published 25 books—including 6 encyclopedias—and more than 2,000 articles in over 60 publications worldwide. He is a frequent speaker at industry trade shows, association meetings, and customer events. He lives in Sterling, Virginia, and can be reached via e-mail at nmuller@ascent-llc.com.